FOODBORNE ILLNESS

A
LANCET
REVIEW

FOODBORNE ILLNESS

Editorial Advisers: W. M. Waites and
J. P. Arbuthnott

Edward Arnold
A division of Hodder & Stoughton
LONDON   MELBOURNE   AUCKLAND

© 1991 Edward Arnold

First published in Great Britain 1991

British Library Cataloguing in Publication Data

Foodborne illness. – (A Lancet review)
   I. Series
   363.19

   ISBN 0–340–55570–X

Photoset in 9/10pt. Plantin by Anneset, Weston-super-Mare, Avon
Printed in Great Britain for Edward Arnold, a division of Hodder
& Stoughton Limited, Mill Road, Dunton Green, Sevenoaks, Kent
TN13 2YA by St Edmundsbury Press Ltd, Bury St Edmunds, Suffolk
and bound by Hartnolls Ltd, Bodmin, Cornwall.

Foreword

A the turn of the century, Adolphe Smith arrived at "Packing Town", Chicago, USA, with a commission from *The Lancet* to report on the sanitary conditions of the world's largest meat market. His description of the surroundings and working conditions is extraordinary. He wrote "There are innumerable rafters, sharp angles, nooks, and corners where blood, the splashing of offal, and the sputum of tuberculous workers can accumulate for weeks, months, and years". Stored meat would often become covered with the faeces of rats, and workers had to eat amidst the general filth. Moreover, Smith noted, "Close at hand there are closets and they are in some places only a few feet from the food." It was not unusual for a worker to fall into the rendering vats.

A little earlier, on the other side of the Atlantic, the Lancet Sanitary Commission in Great Britain uncovered some equally horrifying practices. Food was commonly adulterated—alum was added to flour, red lead to cayenne pepper, and copper arsenate to confectionery. There was a roaring trade in cheap offal and in old and diseased meat. Bakers would wash in the water used for the next batch of dough, and milk was diluted with water from the cattle trough or farmyard pump.

By comparison, today's horror-stories about food seem tame. Nonetheless, foodborne illness is a serious public health issue world wide; costs, in terms of human illness and economic loss, can be immense. The reviews in this book, first published in *The Lancet* in 1990, explain the causes and extent of foodborne illness, and what can be done about it. The twenty-two contributors take us on a journey along the food chain. In the *Lancet* office, much of the inspiration came from my colleague Pia Pini; our guides, Will Waites and John Arbuthnott, who provided invaluable advice and time, lead the way.

ROBIN FOX
The Lancet

Contributors

C Adams, PhD
Food Safety and Inspection Service, US Department of Agriculture, Washington, DC 20250, USA

H Appleton, PhD
PHLS Virus Reference Laboratory, Central Public Health Laboratory, 61 Colindale Avenue, London NW9 5HT, UK

J P Arbuthnott, ScD
Department of Medical Microbiology, Queen's Medical Centre, Nottingham NG7 2UH, UK

D L Archer, PhD
Center for Food Safety and Applied Nutrition, Food and Drug Administration, Washington, DC 20204, USA

A C Baird-Parker, PhD
Unilever Research, Colworth Laboratory, Sharnbrook, Bedford MK44 1LQ, UK

D P Casemore, PhD
Public Health Laboratory, Glan Clwyd Hospital, Bodelwyddan, Rhyl, Clwyd LL18 5UJ, UK.

J G Collee, FRCPath
Department of Medical Microbiology, University Medical School, Teviot Place, Edinburgh EH8 9AG, UK.

E Mary Cooke, MD
Public Health Laboratory Service, 61 Colindale Avenue, London NW9 5DF, UK.

M P Doyle, PhD
Department of Food Science, Georgia Experiment Station, University of Georgia, Griffin, Georgia 30223, USA.

G R Fenwick, PhD
AFRC Institute of Food Research, Norwich Laboratory, Colney Lane, Norwich NR4 7UA, UK.

A M Johnston, MRCVS
Royal Veterinary College, University of London, Hawkshead Lane, North Mimms, Hatfield, Hertfordshire AL9 7TA, UK.

D Jones, PhD
University of Leicester, Dept. of Microbiology, Medical Sciences Building, University Road, Leicester LE1 9HN, UK.

B M Lund, PhD
AFRC Institute of Food Research, Norwich Laboratory, Colney Lane, Norwich NR4 7AU, UK.

M R A Morgan, PhD
AFRC Institute of Food Research, Norwich Laboratory, Colney Lane, Norwich NR4 7UA, UK.

D Roberts, PhD
PHLS Food Hygiene Laboratory, Central Public Health Laboratory, 61 Colindale Avenue, London NW9 5HT, UK.

C J Ryder, MSc
Microbiological Safety of Food Division, Ministry of Agriculture, Fisheries and Food, Ergon House, c/o Nobel House, 17 Smith Square, London SW1P 3JR, UK.

P A Salsbury, BSc
Center for Food Safety and Applied Nutrition, Food and Drug Administration, Washington, DC 20204, USA.

M B Skirrow, FRCPath
Public Health Laboratory, Gloucestershire Royal Hospital, Gloucester GL1 3NN, UK.

P Thompson, BA
Center for Food Safety and Applied Nutrition, Food and Drug Administration, Washington, DC 20204, USA.

E Todd, PhD
Bureau of Microbial Hazards, Health Protection Branch, Ottawa K1A OL2, Ontario, Canada.

H S Tranter, PhD
Division of Biologics, PHLS Centre for Applied Microbiology and Research, Porton Down, Salisbury, Wiltshire SP4 0JG, UK.

W M Waites, PhD
Department of Applied Biochemistry and Food Science, Nottingham University, Sutton Bonington, Loughborough LE12 5RD, UK.

Contents

1
Foodborne illness: an overview

W. M. Waites and J. P. Arbuthnott

Food is a basic human need. To meet this requirement, food production systems are increasing in complexity and size throughout the world. For example, in the European Community total turnover in the food industry is 400 000 million ecu (17% of all manufacturing turnover): 180 000 food-manufacturing businesses employ 2·5 million people (11% of all manufacturing jobs). Foodborne illness is a major public health concern world wide and cuts across national boundaries: in terms of human illness and economic loss the costs can be immense. Controversy about the safety of food has had a huge impact on the food industry and the farming community in many countries. Unsafe food has to be withdrawn from sale and destroyed, and adverse publicity can lead to further economic loss, closure, law suits, and prosecution. Loss of confidence in a certain food or producer can be long term. For example, sales of corned beef after the 1964 typhoid outbreak in Aberdeen, Scotland, were not restored to pre-outbreak levels for about 20 years.

The aim of advisory bodies and governments must be to clarify food safety issues and to establish effective guidelines for control of production, processing, distribution, and sale of food. Additionally, clear guidance on hygienic handling and cooking of food in catering establishments and in the home is needed. The main concern must be the presence of potentially pathogenic microbes and their toxins in food. It is unrealistic to expect that such pathogens, which are widespread in the environment, animals, and man, can be excluded completely from all stages of the food chain; however, this does not remove the obligation to ensure that food is of the highest microbiological quality and is safe for human consumption.

Food safety encompasses a large and diverse area of agriculture, food science and technology, manufacture and processing, as well as microbiology, epidemiology, and human and veterinary medicine. Despite much effort over many years, several fundamental matters continue to be hotly debated—eg, definitions of foodborne disease and food-poisoning, sources of infection, incidence of disease, and the relative importance of particular pathogens and toxic chemicals. This book contains 17 reviews that focus attention on what is known about the principal agents and sources of foodborne illness. Illness due to contaminated food and drink (by microorganisms or their toxins or by chemicals) is usually characterised by diarrhoea or vomiting, or both, but also includes other conditions, such as listeriosis and botulism, which involve parts of the body other than the alimentary tract. It is difficult to be precise about the extent of foodborne illness in the community and to assess the risk of becoming ill from eating contaminated food. It is much easier to assess risks associated with other human activities such as smoking, drinking alcohol, driving cars and motorbikes, taking part in sporting activities, or even crossing the road.

In the UK, the complex system of gathering information about the incidence of foodborne illness is probably the most comprehensive in Europe and it is interesting to contrast the thorough UK reporting systems with the rather patchy situation that prevails in North America. In England and Wales, the Public Health Laboratory Service (PHLS) has a central role through the 52 area public health laboratories, the Communicable Disease Surveillance Centre, and the reference centres at the Central Public Health Laboratory. There are similar, but not identical, surveillance systems in Scotland and Northern Ireland.[1]

Analysis of trends revealed by such surveillance forms the basis of much of our knowledge of the epidemiology of foodborne disease, and there is increasing awareness of the need to match information from human surveillance with that obtained on zoonotic diseases in animals.

Surveillance of foodborne diseases based on such systems not only is vital for evaluation of trends but also is particularly useful in providing early warnings of local or generalised outbreaks, as in the present UK epidemic of *Salmonella enteritidis* PT4 which began after 1984. The food sources responsible can sometimes be identified and appropriate measures taken to control the outbreak. Epidemiological data also form the grounds for the introduction and subsequent revision of regulatory and administrative arrangements for food safety, public health, and animal health. For example, data presented to the committee chaired by Sir Mark Richmond set up by the UK Government to investigate the microbiological safety of food have led to the introduction of new regulatory and administrative changes. In sporadic foodborne disease, such as campylobacteriosis, and when the pathogenesis of the disease is poorly understood (eg, listeriosis), better surveillance systems are needed. There is room for improvement in the gathering and publication of data about man and animals. Additionally, specific surveillance studies to answer questions about causes and control of foodborne disease need to be done.

The food chain

In less developed countries, up to 30% of the food produced is spoilt before it can be consumed either by insect infestations or by microbial growth. The need to obtain enough to eat and drink often outweighs concern about foodborne and waterborne illness; in such circumstances, the incidence of waterborne illness is much higher. Developed countries generally maintain food quality and safety by use of a highly integrated and sophisticated system of food production, processing, transportation, storage, and retailing. Thus, food is often consumed long after and some distance away from the point of primary production. The monetary value added to food by processing, manufacturing, and distribution can be more than three times that of farm output (Fig 1.1). The sophistication of manufacturing and processing techniques has led to the centralisation of such activities into larger units so that a mistake made at a critical point in the food chain can lead to the survival of a toxin or survival and subsequent growth of a pathogenic organism, which will then be spread into numerous products distributed over a wide geographic area. The identification of hazards associated with processes, sites, or raw materials, and their monitoring and control by the use of the hazard analysis critical control point system (HACCP)[2] is urgently needed not only in food manufacture and processing, but throughout the food chain.

Once a toxin or pathogen gets into the food chain it is difficult to remove. Some of the best examples are environmental contaminants, such as polybrominated biphenyls (PBBs) in animal feed used on farms in southern Michigan, USA;[3,4] radioisotope contamination scattered across Europe from the nuclear installation at Chernobyl; contaminated cooking oil in Spain; and lead in animal feed in the Netherlands and the UK. The problems and costs of contamination by bacteria can be even greater because (apart from *Campylobacter jejuni*) they can multiply in food. Recent cost estimates for all foodborne illness in the USA range from US$7700 to US$23 000 million; bacterial foodborne illness accounts for about 80% and salmonellosis about 47% of this total.[5] Costs to individual production units can also be high. The outbreak of listeriosis in California, USA, in 1985 (142 cases, 47 deaths), which was associated with a "Mexican-style" soft cheese, has reputedly cost the company (including litigation) more than $700 million. The salmonellosis outbreak in Cumbria, England, in 1985 (76 cases, 1 infant death) due to consumption of infant dried milk resulted in a more than £22 million loss to the company and one of the two production units was subsequently closed with the loss of more than 100 jobs.[6]

CURRENT TRENDS

Eating habits, especially in developed countries, are changing rapidly. There is a public demand to reduce the use of "chemicals" at all stages of the food chain—from primary production (pesticides, herbicides, and antibiotics) to the finished product (preservatives). Therefore, to produce more "natural" foods with better flavour, processing is reduced, and to maintain safety and shelf life, the cold storage and distribution chain has been extended. The introduction of novel foods (eg, mycoprotein), new composite foods, and processes (microwaving, sous vide, and irradiation) should be considered with care. In developed countries there is an increasing tendency for people to eat more meals outside the home, to buy food in bulk (and less often), and to eat more "fast" foods, which require less preparation. Also, more people intentionally restrict their diet for reasons either of health or of conscience. Vegetarians, for example, are exposed to greater amounts of natural toxicants and mycotoxins in plant materials. Risk can be increased even by a simple dietary change, such as greater consumption of potato skins (as in baked potatoes), which leads to greater exposure to glycoalkaloids and thus to possible chronic effects due to the potent gut-permeabilising activity of these compounds.[7]

Generally, risks associated with such long-term contact with intoxicants are often difficult to quantify. The assessment of risks which result in the immediate or rapid onset of foodborne illness is much easier. The possibility that bovine spongiform encephalopathy, which has a long incubation period, could be transmitted to man remains hypothetical and is especially difficult to assess. There is, however, no evidence of such transmission. The effects of other pathogenic microorganisms within the food chain are less speculative. That man and other animals can be carriers of various pathogenic micro-organisms without overt features of disease is well known. Such animal carriers are particularly important sources of human illness (eg, salmonellosis due to *S enteritidis* infection in poultry in both Europe and the USA), but human carriers have a causal role in only about 4% of reported foodborne illness. Although the relation of foodborne salmonellosis with pets and animals used for food is complicated, it has been apparent for many years[8] that the

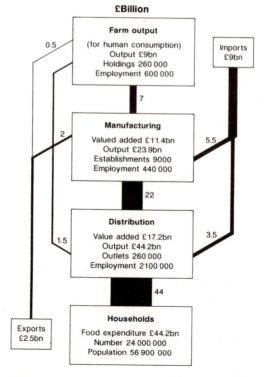

£Billion

Farm output
(for human consumption)
Output £9bn
Holdings 260 000
Employment 600 000

0.5

Imports
£9bn

7

Manufacturing
Valued added £11.4bn
Output £23.9bn
Establishments 9000
Employment 440 000

2

5.5

22

Distribution
Value added £17.2bn
Output £44.2bn
Outlets 260 000
Employment 2100 000

1.5

3.5

44

Exports
£2.5bn

Households
Food expenditure £44.2bn
Number 24 000 000
Population 56 900 000

Billion=thousand million
Source=Economics and Statistics (Food) Division, Ministry of Agriculture, Fisheries and Food.
Figure 1.1　UK food chain 1986

arrangement for transport and slaughter of animals used for food increases the proportion of animals carrying *Salmonella*—eg, from 0·5% of calves (UK) and 7% of pigs (USA) leaving the farm to 36% and 50%, respectively, after slaughter.

Any one part of the food chain is clearly integrated very closely with others. Division in control and surveillance is likely to lead to failure because of the lack of understanding of problems which arise from a new process or a "new" pathogen. Although *S enteritidis* PT4 might be regarded as the best example of such a pathogen, campylobacter has now become the most commonly reported cause of foodborne infection, at least in England and Wales (with an estimated annual cost of nearly £90 million) and in parts of the USA.[9]

The complexity of the food chain and the possibility that a change in one process in one part can have a detrimental effect in a different part has been given an added emphasis by the increase in incidence of listeriosis. Since 1983, major outbreaks of foodborne listeriosis have occurred in the USA (twice) and Switzerland (more than 200 cases and 91 deaths) but not in the UK, where only 4 cases have been attributed to food. However, even in England and Wales, the number of cases increased until 1988; this may have been due to an increase in the prevalence of *Listeria monocytogenes* in the environment, partly due to changes in methods of silage production;[10] there is also increased awareness by food manufacturers of the importance of *L monocytogenes*, especially when the organism is present in chilled areas of food preparation and storage. The most recent data show a substantial decrease in reported cases of listeriosis since 1988.

Other bacteria are also emerging as important agents of foodborne disease. In Belgium, Canada, the Netherlands, Australia, and parts of Germany, *Yersinia enterocolitica* rivals *Salmonella* as a cause of acute gastroenteritis. Like *L monocytogenes*, *Y enterocolitica* can grow at temperatures as low as 0°C and the increased use of refrigeration in the food chain may play a part in the increase in prevalence of this organism in food. In addition to changes in processing, dietary differences are important in determining the organisms that are principally responsible for foodborne disease in a certain country. For example, gastroenteritis due to *Vibrio parahaemolyticus* is nearly always associated with consumption of contaminated seafood.[11] Although recognised as a health hazard in seafood consumed world wide, *V parahaemolyticus* is the principal cause of foodborne disease only in countries such as Japan, where raw fish is widely consumed; moreover, the organism is rarely isolated when the seawater temperature falls below 13°C. Other vibrios (*V fluvialis*, *V hollisae*, *V mimicus*, and *V furnissii*) are also associated with gastroenteritis,[12] though less often. Primary septicaemia due to *V vulnificus* infrequently develops after consumption of raw seafood; however, the mortality rate is high (>60%). In developing countries, enterotoxigenic and enteropathogenic *Escherichia coli* are an important cause of gastrointestinal disease (the reservoirs of infection are mainly infected foodhandlers and water contaminated with sewage). By contrast, in the USA the principal reservoir of enterohaemorrhagic *E coli* O157:H7 is the intestinal tract of dairy cattle.

Various gram-positive bacteria produce illness that is usually due not to the organism itself but to production of a preformed toxin in the food—eg, *Staphylococcus aureus*, *Clostridium botulinum*, and several *Bacillus* species. Correct processing, storage, and foodhandling may be even more important with these organisms. Staphylococcal food-poisoning is the only staphylococcal disease that is not associated with growth of the organism in or on the human body; about 100–200 ng of the enterotoxin will produce illness and because this toxin is heat resistant (unlike the organism), cooking will not render food safe. It is therefore necessary to avoid contamination with toxin-producing bacteria; refrigeration is the best way to prevent formation of toxin. Staphylococcal food-poisoning is, however, rarely fatal. By contrast, botulism is a very serious illness; as little as 0·1 g of food in which *Cl botulinum* has grown can cause the disease and as many as 20% of cases of botulism result in death. Like *Cl botulinum*, *B cereus* produces heat and irradiation resistant endospores, and, because of their presence in foods, heat processing has to be severe enough to kill spores or prevent their outgrowth (121°C for 20 min in canning, or 135°C for 5 s in ultra high temperature processing). *Cl botulinum* toxin and endospores also survive the levels of irradiation (10 kGy) currently proposed for treatment of food, so that strict control during food processing will be needed if the procedure is to be widely used in the food industry. Another endospore forming bacterium, *Cl perfringens*, generally causes illness by the formation of enterotoxin during spore formation in the gut. Symptoms appear only after ingestion of heavily contaminated food (>100 million bacteria/g). Since the organism can double in numbers every 7·1 min at 41°C,[13] the importance of correct foodhandling becomes obvious.

Control

Improved control can be achieved by better microbiological quality of raw materials, by prevention of cross-contamination throughout the food chain, and by better handling and adequate storage and cooking of food. Thus, there must be adequate controls at all points in the food chain at which microbial and toxicological hazards can occur. For example, during canning or other heat processing (such as ultra high temperature treatment, pasteurisation, or cooking), the maintenance of the required temperature for sufficient time to inactivate microorganisms and their toxins is vital. Likewise, the increase in acidity during cheese or yoghurt manufacture is needed not only to provide an acceptable product with the correct taste and "mouth feel", but also to ensure conditions under which pathogenic or toxigenic bacteria cannot grow or produce toxin. The monitoring of such controls and use of HACCP are, or should be, incorporated into good manufacturing practice for the food and drink industry in all countries.[14] It is the responsibility of managers that all staff are adequately trained in basic food hygiene. Viruses and protozoa do not multiply or produce toxins in food; food (or water) merely acts as a vehicle for their transfer. Viruses are difficult to detect in food; nonetheless, various viruses from the gastrointestinal tract, such as hepatitis A and small round structured viruses of the Norwalk group, are increasingly acknowledged as agents of foodborne illness. These viruses are especially important in shellfish grown in water contaminated with faecal microorganisms, although person-to-person transmission is a more common route of infection. Confirmed foodborne protozoal infection is rare in developed countries; however, there is growing evidence of transmission by means of potable water—eg, *Cryptosporidium* and *Giardia*. Standards recommended by the European Commission will undoubtedly lead to tighter controls over the levels of such potential pathogens in potable water in Europe.

The future

In general, despite the capacity of microorganisms to evolve and adapt, the difficulties associated with foodborne illness can be overcome provided that our monitoring systems give us an early warning, that our basic science gives us sufficient understanding of the organism or the toxin, and that our control and educational systems can react quickly enough. Unfortunately, our fundamental understanding of the interaction of microorganisms with the environment, of their ability to survive and grow in extreme environments, and of the differences between closely related strains, which allow only some to survive food processing, is inadequate. In view of the financial and human cost of foodborne illness, such an understanding will obviously benefit both the community and the food industry; the rewards will greatly outweigh the costs of such research.

References

1. The microbiological safety of food. Part II. Report of the Committee on Microbiological Safety of Food. London: HM Stationery Office, 1991: 24–52.
2. Microorganisms in food. Vol 4. Application of the Hazard Analysis Critical Control Point (HACCP) system to ensure microbiological safety and quality. Oxford: Blackwell Scientific, 1988.
3. Munro IC, Charbonneau SM. Environmental contaminants. In: Roberts HR, ed. Food safety. New York. Wiley Interscience, 1981: 141–80.
4. Office of Technology Assessment. Environmental contaminants in food. Washington, DC: US Government Printing Office, 1979.
5. Todd ECD. Preliminary estimates of costs of foodborne disease in the United States. *J Food Protect* 1989; **52**: 595–601.
6. Waites WM. The magnitude of the problems. In: Miller FA, ed. Food safety in the human food chain. Reading: Centre for Agricultural Strategy Paper 20, Reading University, 1990: 27–38.
7. Gee JM, Price KR, Ridout CL, Johnson IT, Fenwick GR. Effects of some purified saponins on transmural potential difference in ammalian small intestine *Toxical in vitro* 1989, **3**: 85
8. Ingram M. Meat preservation—past, present and future. *R Soc Health J* 1972; **92**: 121–30.
9. MacDonald KL, O'Leary MJ, Cohen ML, et al. *Escherichia coli* O157:H7, an emerging gastro-intestinal pathogen. *JAMA* 1988; **259**: 3567–70.
10. Fenlon DR. With birds and silage as reservoirs of *Listeria* in the agricultural environment. *J Appl Bacteriol* 1985; **59**: 537–43.
11. West PA. The human pathogenic vibrios: a public health update with environmental perspectives. *Epidemiol Infect* 1989; **103**: 1–34.

12. Anon. Shuck your oysters with care. *Lancet* 1990; **336**: 215–16.
13. Willardsen RR, Busta FF, Allen CE, Smith LB. Growth and survival of *Clostridium perfringens* during constantly rising temperatures. *J Food Sci* 1978; **43**: 470–75.
14. Anon. Food and Drink Manufacture—Good Manufacturing Practice: a guide to its responsible management. London: Institute of Food Science and Technology (UK), 1988.

2
Epidemiology of foodborne illness: North America

Ewen Todd

Foodborne disease surveillance began in the USA at the beginning of this century and the first compilations of milkborne outbreaks were prepared in 1923. From 1938 onwards more complete data, which covered foodborne and waterborne outbreaks, were collected. The association of infant diarrhoea and typhoid fever with milk and water stimulated health authorities to introduce mandatory pasteurisation of milk and chlorination of municipal water supplies much earlier than in most other parts of the world. Annual summaries of foodborne disease outbreaks were produced by the Centers for Disease Control (CDC) from 1966 to 1981 as individual publications, but from 1982 to 1987 these were included as supplements of the *Morbidity and Mortality Weekly Report*. A standard two-page form is used by each state to record data about date and place of outbreak, aetiological agent, food implicated, foodhandling errors that contribute to outbreaks, and place of mishandling. Up to 1978, complete line listings for each of the outbreaks were published, but because of economic restraints only those of known aetiology were recorded for 1979; then they ceased to be published at all. There is a certain variability in data reported to the Center for Infectious Diseases, CDC, because states are encouraged, but not obliged, to supply outbreak data. According to Bartleson[1] "although some states and territorial health agencies have admirably collected and reported foodborne illness, other jurisdictions have exhibited an abysmal, noncooperative lack of concern for foodborne disease investigation and reporting". Some states report rates 200 times those of others. The quality of reports also varies but has slowly improved over the past decade.[1] From the latest available data (1983–87)[2] fewer outbreaks but more cases seem to be reported, which indicates perhaps that more effort is being put into the larger outbreaks; information about small incidents may never reach the national data base.

The Canadian Food-borne Disease Reporting Centre of the Health Protection Branch began publishing information of all foodborne and waterborne disease incidents in 1973. The style of presentation is based on the more complete CDC publications, with line listings occupying between 30 and 65% of the pages. The extent of this detail is justified because this is what the reader most often wants to see—ie, complete information about specific outbreaks with the cause described, if known. Succinct forms of these summaries are also published in the *Journal of Food Protection* with examples of unusual incidents that happen each year. Data up to 1983 and 1984 have been published, with 1985 and 1986 summaries available in 1990.

Critics have argued that the value of the data is reduced because they are published much too late after the events have occurred. Three reasons are often given. First, complete data from large federated countries with at least three levels of government are difficult to obtain (local/regional/county: state/provincial: federal); even at the federal level in the USA only the more important incidents investigated by the Food and Drug Administration are included in the CDC data. Second foodborne disease has never been given high priority in public health, and resources are limited. Third, knowledge about the long-term trends is often more important for making decisions about how foodborne disease can be reduced and, therefore, delays are not always critical.

The Canadian data are more complete than the US data because each province and territory reports annually. For the years 1978–82, 14 340 cases of foodborne disease were reported in the USA compared with 6190 in Canada—less than a three-fold difference in numbers, compared with a ten-fold difference in population. Of course, collection is easier from 12 jurisdictions than from 55.

Estimated cases and costs

These foodborne disease reports in which food is implicated should not be confused with reports of notifiable diseases and isolations of enteric pathogens which have been collected at the national level for many years by both countries. Notifiable diseases are obligatory notifications of specific diseases sent by physicians directly to public health authorities (eg, typhoid, salmonellosis, shigellosis, botulism); isolations of enteric pathogens, such as *Campylobacter* spp, *Salmonella* serovars, and *Shigella* spp, are based on monthly laboratory reports. Since there is no requirement for an association with food, the number of notifiable diseases or isolations of enteric pathogens is higher than the number of foodborne disease reports. For example, for 1983–86, the mean numbers of cases of samonellosis in the USA were: notifiable diseases, 50 110;[3] laboratory isolates, 43 434;[4] foodborne outbreak associated cases, 7350.[2] The comparable Canadian data from the Health Protection Branch for salmonellosis were 9318, 8846, and (1983–85 only) 2565, respectively. However, the actual number of cases is much higher. Recent estimates of USA salmonellosis cases range from 790 000 to 3690 000 (median 1920 000).[5-9] Calculations have been made for all cases of variously defined foodborne disease; numbers for the USA range from 6·3 million (cases of known cause only) to 81 million (Table 2.1). This huge variation is explained by the different calculation methods and the number of assumptions made. The same difficulty occurs when the annual number of deaths caused by foodborne illness is estimated. For the USA, these range from a conservative figure of 523[9] to 7041,[8] though the latter is regarded as the more realistic. These data show that, even with large variations, foodborne disease is widespread and can have serious health consequences.

Estimates have also been extended to include costs. Most investigators who study costs have used medical expenses, lost human productivity, and sometimes the value of deaths, since these indices are relatively easy to measure in a broad sense. Other costs, such as losses to the food supplier (recall, destruction of the food, extra testing of food samples, reprocessing and repackaging, loss of market) and legal action (fines, lawyers' expenses, payment of damages) are much more difficult to assess and have to be calculated for each incident; nonetheless, these other costs can be more expensive than the medical and human productivity losses, especially when there are problems with processed food.[13] Recent cost estimates for all foodborne illness in the USA range from US$7700 to US$23 000 million (table). Salmonellosis, campylobacteriosis, and staphylococcal intoxication are the most extensive and expensive of the foodborne diseases; each case of salmonellosis costs between US$700 and US$1350.[8,9] Preliminary information for Canada indicates that there are about 2·2 million cases each year (about 1 million of microbiological origin and about 1 million of unknown cause) costing Can$1335 million.[14]

The estimation of numbers of foodborne disease cases and costs is a relatively new discipline and further refinement of methods is likely to give more accurate data: these will probably be based on intensive surveillance for a limited time in several selected locations (sentinel studies), including origin of laboratory-confirmed cases, and on appropriate case-control studies. Even without such data, however, there is general agreement that food-associated illness is unnecessarily high.

Recent foodborne disease concerns

Some of the current problems for the USA and Canada are as follows. Cases of salmonellosis have steadily increased in both countries. For example, notifiable diseases in the USA more than doubled from 22 612 in 1975 to 48 948 in 1988, with a peak in 1985 of 65 347 owing to the *Salm typhimurium* outbreak described below. *Salm typhimurium* is the most frequently isolated serotype and has been responsible for the largest foodborne outbreaks ever reported in both countries: improperly pasteurised milk caused 16 659 laboratory confirmed cases and up to 197 581 estimated cases in Illinois in 1985,[15] and cheddar cheese made from incompletely pasteurised milk in eastern Canada was responsible for 2700 cases in 1984.[16] A more recent concern is the rise in cases of *Salm enteritidis* in New England and in the mid-Atlantic states; there were 140 outbreaks (4976 cases and 30 deaths) between 1985 and 1989, in 65 of which grade A shell eggs were implicated.[17,18] In the USA, eggs from positive egg-laying flocks are allowed to be pasteurised, and vaccines may be used for egg-laying but not for breeding flocks.

Table 2.1   Estimated numbers of US cases and costs per year of foodborne gastroenteritis

Reference	Method	Type of illness	Estimated no of cases	Costs (US$)*
Garthwright et al[10]	Extrapolation of health surveys	Intestinal infectious disease†	99 million	23 000 million
Archer and Kvenberg[11]	Extrapolation of health surveys	Foodborne diarrhoeal disease	24–81 million	High
Kvenberg and Archer[12]	Extrapolation of health surveys	Foodborne diarrhoeal disease	33 million	7700 million
Todd[9]	Extrapolation of cases of specific foodborne diseases	Foodborne disease	12·6 million‡	8400 million
Roberts[8]	No of cases of specific foodborne diseases from Bennett et al[6]	Foodborne disease	6·3 million§	4800 million

*Medical expenses and lost human productivity; Todd[13,14] adds costs to the food supplier, legal expenses, &c.
†Vomiting and diarrhoea without respiratory infection.
‡Cases both of known and unknown agents.
§Cases only of known bacterial agents.

Since 1982, cases of haemorrhagic colitis, haemolytic uraemic syndrome, and thrombotic thrombocytopenic purpura, with deaths in elderly people, have been linked to meat and milk containing *E coli* O157:H7. Outbreaks have occurred in nursing homes, daycare centres, and schools, and have been associated with "fast food" outlets. Sporadic case rates are at their highest in the summer, mainly in western Canada and northwest USA. Undercooked hamburger meat (ground beef) is regarded as a major source of the organism. The incidence rate for *E coli* O157:H7 in the Puget Sound area, Washington state, is 8 per 100 000 person-years compared with 150 for *Campylobacter* and 21 for *Salmonella*.[7] Other verotoxin-producing *E coli* are also suspected of causing human illness but the epidemiological proof is, so far, limited.

Although campylobacteriosis is a common cause of sporadic disease, outbreaks are rare. The numerous sporadic cases may be poultry-associated since many carcasses contain large numbers of campylobacters. In Canada, isolations of campylobacters exceeded those for salmonellae in 1989.

Listeria monocytogenes is a major concern for the food industry because the organism is common in animals and in the environment. The most recent large outbreak of listeriosis (associated with "Mexican-style" soft cheese) in North America was in 1985.[19] Sanitation in processing plants, heat processing, and cold storage are areas where proper control must be established to reduce the prevalence of the organism. In addition, laboratory markers to differentiate virulent from less virulent strains are needed to improve epidemiological studies of listeriosis.

Infections with *Vibrio* spp from shellfish are reported from the warmer southern regions of the USA, where the organisms naturally occur in brackish seawater, and also from regions in the USA where these shellfish are distributed for sale. The most serious of these infections is due to *V vulnificus* which can survive in oysters for up to 14 days at refrigeration temperatures. CDC recommends that people at risk (ie, those with liver disease) should not eat raw shellfish because the risk of septicaemia and death is high.[20]

Staphylococcal outbreaks are common but rarely involve commercially prepared food. In 1989, more than 100 cases in several states were associated with 100-ounce cans of mushrooms imported from the People's Republic of China.[21] The toxin, which may not be inactivated by cooking or canning temperatures, was produced before canning or reached the product through post-process leakage. Botulism, which is not uncommon in the arctic regions, arises from consumption of traditionally prepared marine mammal and fish products. In 1985 and 1986, outbreaks in Vancouver restaurants were associated with unrefrigerated garlic in oil, and bottled mushrooms, respectively[22,23] *Kapchunka* (ungutted, salt-cured, partly-dried whitefish) has caused enough cases of botulism in the past decade to prevent its sale in the USA. Although there are occasional cases of paralytic shellfish poisoning from both west and east coasts, illness from amnesic shellfish poisoning was documented for the first time in the world from mussels harvested in Prince Edward Island in 1987.[25] Domoic acid, an excitotoxic aminoacid responsible for the gastroenteritis and brain damage, originated from the diatom *Nitzschia pungens*.

Very recently there has been concern over the use of L-tryptophan-containing products (LTCP) used as a dietary supplement, which cause eosinophilia-myalgia syndrome (EMS).[26,27] There have been more than 1500 cases and 20 deaths in the USA and 10 cases in Canada. There are fewer Canadian cases and deaths because LTCP are available only through prescription in that country. A very high proportion of the cases had consumed LTCP before getting myalgia and arthralgia, followed by dyspnoea, cough, rash, and an eosinophil count of >1000 cells/μl. In severely affected patients, progressive polyneuropathy developed, similar to victims of the 1981 Spanish toxic oil syndrome. There are two likely contaminants, both derivatives of L-tryptophan, which may have been produced by a genetically modified strain of *Bacillus amyloliquefaciens*, the organism used in the industrial biosynthesis of the aminoacid, but there is still no proof that the genetic modification process was responsible for the presence of the contaminants.[28]

To improve control of foodborne diseases there needs to be enhanced investigation of outbreaks through better coordination of health agencies at different jurisdictional levels (including regular dissemination of material by means of effective reporting forms); better trained medical personnel, laboratory staff, and health inspectors; and computer networking. More information on factors that contribute to outbreaks should be stressed, so that proper control measures based on the HACCP (hazard analysis critical control point) principle can be introduced.[1]

References

1. Bartleson CA. Foodborne disease surveillance. In: Felix CW, ed. Food protection technology. Chelsea, Michigan: Lewis Publishers, 1987: 141–55.
2. Bean NH, Griffin PM, Goulding JS, Ivey CB. Foodborne disease outbreaks, 5-year summary, 1983–1987. *MMWR* 1990; **39**: 15–57.
3. Centers for Disease Control. Summary of notifiable diseases, United States, 1988. *MMWR* 1988; **37**: 1–58.
4. Hargrett-Bean N, Pavia AT, Tauxe RV. *Salmonella* isolates from humans in the United States, 1984–1986. *MMWR* 1988; **37** (SS-2): 25–31.
5. Chalker RB, Blaser MJ. A review of human salmonellosis: III Magnitude of *Salmonella* infection in the United States. *Rev Infect Dis* 1988; **10**: 111–24.
6. Bennett JV, Holmberg SD, Rogers MF, Solomon SL. Infections and parasitic diseases. In: Amler RW, Dull HB, eds. Closing the gap: the burden of unnecessary illness. New York: Oxford University Press, 1987: 102–14.
7. MacDonald KL, O'Leary MJ, Cohen ML, et al. *Escherichia coli* O157:H7, an emerging gastro-intestinal pathogen. *JAMA* 1988; **259**: 3567–70.
8. Roberts T. Human illness costs of foodborne bacteria. *Am J Ag Econ* 1989; **71**: 468–74.
9. Todd E. Preliminary estimates of costs of foodborne disease in the United States. *J Food Protect* 1989; **52**: 595–601.
10. Garthwright WE, Archer DL, Kvenberg JE. Estimates of incidence and costs of intestinal infectious disease in the United States. *Public Health Rep* 1988; **103**: 107–15.
11. Archer DL, Kvenberg JE. Incidence and cost of foodborne diarrheal disease in the United States. *J Food Protect* 1985; **48**: 887–94.
12. Kvenberg JE, Archer DL. Economic impact of colonization control on foodborne disease. *Food Technol* 1987; **41**: 77–98.
13. Todd E. Economic loss from foodborne disease and non-illness related recalls because of

mishandling by food processors. *J Food Protect* 1985; **48**: 621–33.

14. Todd E. Preliminary estimates of costs of foodborne disease in Canada and costs to reduce salmonellosis. *J Food Protect* 1989; **52**: 586–94.

15. Ryan CA, Nickels MK, Hargrett-Bean NT, et al. Massive outbreak of antimicrobial-resistant salmonellosis traced to pasteurized milk. *JAMA* 1987; **258**: 2369–74.

16. Todd E. Foodborne and waterborne disease in Canada: 1984 Annual summary. *J Food Protect* 1989; **52**: 503–11.

17. Steinert L, Virgil D, Bellemore E, et al. Update: *Salmonella enteritidis* infections and grade A shell eggs: United States, 1989. *MMWR* 1990; **38**: 877–80.

18. St Louis ME, Morse DL, Potter ME, et al. The emergence of grade A eggs as a major source of *Salmonella enteritidis* infections: new implications for the control of salmonellosis. *JAMA* 1988; **259**: 2103–07.

19. Lovett J. *Listeria monocytogenes*. In: Doyle MP, ed. Foodborne bacterial pathogens. New York: Marcel Dekker, 1989: 283–310.

20. Eastaugh J, Shepherd S. Infectious and toxic syndromes from fish and shellfish consumption: a review. *Arch Intern Med* 1989; **149**: 1735–40.

21. Collins RK, Henderson MN, Conwill DE, et al. Multiple outbreaks of staphylococcal food poisoning caused by canned mushrooms. *MMWR* 1989; **38**: 417–18.

22. Hauschild AHW. *Clostridium botulinum*. In: Doyle MP, ed. Foodborne bacterial pathogens. New York: Marcel Dekker, 1989: 111–89.

23. McLean HE, Peck S, Blatherwick FJ, et al. Restaurant-associated botulism from inhouse bottled mushrooms, British Columbia. *Can Dis Weekly Rep* 1987; **13**: 35–36.

24. Federal Register. Salt-cured, air dried, uneviscerated fish; compliance policy guide: availability. United States Government 1988; **53**: 44949–51.

25. Todd E. Amnesic shellfish poisoning: a new seafood toxin syndrome. In: Graneli E, Sunström B, Edler L, Anderson DM, eds. Toxic marine phytoplankton. New York: Elsevier, 1990: 504–08.

26. Eidson M, Voorhees R, Tanuz M, et al. Eosinophilia-myalgia syndrome and L-tryptophan-containing products: New Mexico, Minnesota, Oregon and New York, 1989. *MMWR* 1989; **38**: 785–88.

27. Heikoff L, Ellis K, Garona JE, et al. Clinical spectrum of eosinophilia-myalgia syndrome: California. *MMWR* 1990; **39**: 89–91.

28. Anon. Tryptophan production questions raised, *Biotechnology* 1990; **8**: 992.

3
Epidemiology of foodborne illness: UK

E. Mary Cooke

This review is based primarily on information collected by the Public Health Laboratory Service (PHLS), England and Wales, from various sources. These data include reports from laboratories to the PHLS Communicable Disease Surveillance Centre (CDSC) of organisms that cause gastrointestinal disease and also of organisms that, though they may be foodborne, produce other infections. Additionally, the reference laboratories, particularly the Division of Enteric Pathogens (DEP), the Food Hygiene Laboratory, and the Division of Microbiological Reagents and Quality Control, collect data about organisms sent to them for identification and typing. Clinical cases of food-poisoning are also reported to the Office of Population Censuses and Surveys; the reporting system is based on a clinical diagnosis alone, irrespective of whether there is supporting laboratory data. Clinical cases of infectious intestinal disease (which may not always be foodborne) are reported through the Royal College of General Practitioners Sentinel Practice scheme; this scheme involves 60 general practices with about 425 000 patients. Finally, special studies and surveys may be done in areas of current interest. These various sources of information are analysed by the CDSC to provide an overview of communicable diseases in England and Wales.

Surveillance in Scotland is the responsibility of the Communicable Diseases (Scotland) Unit, and there is much interchange of information between that unit and the PHLS. In Northern Ireland, the Department of Health and Social Services Regional Information Board collates information based on laboratory reports and statutory notifications. It should be remembered that, especially for mild and common illnesses, there is substantial under-reporting. Individuals with transient diarrhoea may not consult their family doctor, and even if they do the illness may not be investigated. The information that we have, therefore, indicates changes in patterns of foodborne disease but not its true incidence.

Nonetheless, the trends for many foodborne diseases are sufficiently pronounced for us to say which diseases are of increasing public health importance. Three organisms stand out prominently. Salmonellas are becoming increasingly important (Fig 3.1), especially *Salmonella enteritidis* phage type 4 (PT4), which is associated with poultry and eggs, and which is now the most prevalent serotype in man in England and Wales. The total number of salmonella isolates sent to the DEP for typing in 1979 was 10 276; in 1989, 29 998 isolates were received, of which 12 931 were *S enteritidis* PT4. Although they attract less public attention because the associated disease is usually mild and sporadic, campylobacters are another major concern; they continue to account for more identified cases of foodborne illness than do salmonellas (Fig 3.1). Laboratory notifications to CDSC were 8514 in 1979 and provisionally 32 359 in 1989. Listeriosis is thought to be important because, although the disease is rare, reports to the CDSC increased from 70 in 1979 to 251 (provisional) in 1989; because listeriosis is a severe disease; and because of its association with fetal and neonatal infections with high mortality.

Bacterial foodborne disease

SALMONELLA INFECTIONS

Animals are the main source of infection, although spread from person to person may also occur. Infection may be subclinical or cause a serious debilitating or even fatal illness—for example, in elderly people. The route of infection is usually from contaminated meat but spread to other foods may occur during their preparation. The monitoring of salmonella infections is aided by means of specialised methods for characterising the organisms based on the antigenic structure and on subdivision within serotypes by phage typing and other typing methods.[1] Since data were first collected in the 1940s, there have been four striking increases in the number of cases of human salmonellosis in the UK. The first peak (in which *S typhimurium* predominated), in the 1950s, was attributed to infection in cattle and poultry. The second peak, during the late 1960s and the 1970s, was due to serotypes other than *S typhimurium*; there were also peaks of previously unrecorded serotypes, such as *S agona* and *S hadar*, which were probably attributable to contaminated imported animal feed and were associated with infection in poultry. *S typhimurium* was again the main cause of the third rise in salmonellosis in the late 1970s and early 1980s. These infections were attributed to cattle. The most recent increase in salmonellosis, since 1985, is attributable to a huge rise in *S enteritidis* (mostly PT4) cases. Between 1981 and 1988, *S typhimurium* increased less than two-fold and other serotypes were almost unchanged. By contrast, in Scotland there was a similar proportionate increase in *S enteritidis* PT4, but the prevalence of disease caused by other serotypes declined, so that the number of laboratory reports of infection in 1988 was almost the same as that in 1981. The increase in *S enteritidis* PT4 salmonellosis in Northern Ireland was less substantial than that in England and Wales. The reasons for these geographical differences

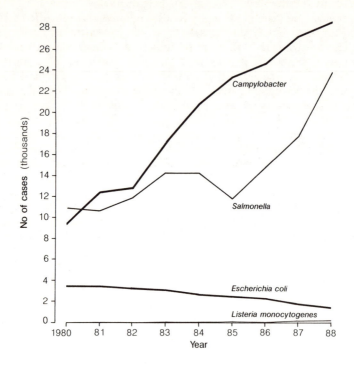

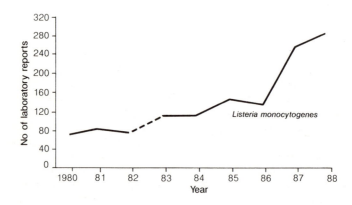

Figure 3.1   Cases of gastrointestinal infection due to *Campylobacter*, *Salmonella*, and *E coli* (EPEC)*, and systemic *L monocytogenes*,† infections, England and Wales, 1980–88.

---

*In children <3 years
†Expanded version shown below: 1980–82, CDSC data only; from 1983, data from both CDSC and PHLS Division of Microbiological Reagents and Quality Control.

are not known. The epidemic of salmonellosis due to *S enteritidis* PT4 has been attributed largely to poultry and hens' eggs,[2] as evidenced by infection in poultry broiler and egg laying flocks, infection of ovarian tissue in poultry, and an increase in the number of outbreaks of human infection associated with hens' eggs.

Prevention of salmonella food-poisoning is a challenge to public health. There are two possible approaches—by attempts to reduce infection in animals used for food and by the subsequent handling of the animal products so that the risks are reduced. The first approach is difficult and involves the treatment of animal feed to remove salmonellas, and the development of salmonella-free breeding stock. An example of such an approach is the policy, introduced by the Ministry of Agriculture, Fisheries and Food in 1989, of slaughtering (poultry) flocks that are infected with *S enteritidis*. There is not yet enough evidence to indicate how successful the measure will be, although the rate of increase of human *S enteritidis* infections in the UK began to decline in 1989. However, during the first half of 1990, the number of human cases of *S enteritidis* infection has continued to increase: 9430 salmonella isolates (4837 PT4) were received by the DEP for identification compared with 9270 (4089) for the same period in 1989. The second approach involves careful handling of animal carcasses to reduce spread of the organism, hygienic precautions in food preparation to prevent cross-contamination and multiplication of the organism, and thorough cooking of all food likely to be affected. Prevention requires attention from a wide variety of people, including animal breeders, veterinarians, epidemiologists, microbiologists, caterers, and also household cooks. The solution is not likely to be easy to achieve. However, a substantial impetus has been given by the reports of the Committee on the Microbiological Safety of Food (Richmond Committee).[4,5] These reports contain wide ranging conclusions and recommendations for the investigation, prevention, and control of foodborne disease.

Typhoid and paratyphoid fever are characteristically infections of man with transmission via food or water or, where hygiene is poor, directly from person to person. Thus, in the UK most infections are acquired abroad; only 296 laboratory reported cases of enteric fever were recorded in 1988. Prevention of enteric fever for UK residents is mainly based on appropriate precautions when travelling abroad.

CAMPYLOBACTERS

Campylobacters are the commonest cause of infectious diarrhoea in England and Wales, with up to 30 000 cases recognised each year.[6] The incidence of the disease is increasing; laboratory reports to the CDSC show a continuous rise since reporting began in 1977 (see also Fig 3.1) and a similar trend is seen in Scotland and Northern Ireland. Although some of this increase is due to the introduction of better laboratory techniques and to changes in the investigation of cases and in reporting practices, it is probable that there has been a real increase in incidence of infection. The organism is found in the gut of animals, both wild and domestic, and infected poultry is an important source; contact with pets has sometimes been noted. Campylobacters do not multiply in foods and most cases are sporadic, although there have been outbreaks attributable to contaminated milk and water supplies. The incidence of this disease could be reduced by control of the infection in broiler chickens, the adoption of hygienic food preparation methods, adequate cooking, and the efficient pasteurisation of milk to prevent milkborne outbreaks. In Scotland, where compulsory pasteurisation was introduced in 1983, milkborne outbreaks have been controlled. The sale of raw milk is still permitted in England and Wales.

LISTERIA

In most cases of listeriosis, proof that the disease is associated with consumption of contaminated food is lacking, although there is much indirect evidence. Outbreaks attributed to contaminated food have been reported from other countries. Epidemiological studies are aided by phage typing and serotyping systems; one serotype, 4bx/4by, is especially common in human cases in the UK. Many foods have been shown to be contaminated with the organism but there have been only 4 single cases of infection that have been directly attributed to a food vehicle in the UK. Listeria is especially troublesome because it can grow, albeit slowly, at refrigeration temperatures. Further evidence of an association with food awaits the completion of a national case-control study that began in July, 1990.

The symptoms of the disease range from a mild influenza-like illness to a severe septicaemia, particularly at the extremes of age and in the immunosuppressed; infection during pregnancy can result in abortion, stillbirth, or birth of a severely affected baby.

There has been a striking increase in reported cases of listeriosis in England and Wales—from about 25 in the early 1970s to 291 in 1988, including 52 deaths and 11 abortions. The number of cases nearly doubled between 1986 and 1987; the rise continued, but more slowly, in 1988 (Fig 3.1). The upward trend, however, did not continue in 1989 when 251 cases (provisional) were recorded. In the first six months of 1990, there were 59 cases of listerosis compared with 152 for the same period in 1989.

Despite the lack of firm evidence of a direct association between sporadic cases, which form the majority, and contaminated food, it is prudent for pregnant women to avoid food known to be sometimes heavily contaminated with *L monocytogenes* (eg, soft cheeses and pâté).

It is noteworthy that small outbreaks, which are not food related, occasionally occur in neonatal units; these can be prevented by standard hospital infection control techniques, particularly handwashing after handling babies.[7-9]

ESCHERICHIA COLI

Although *E coli* normally inhabits the gastrointestinal tract of man and animals, some strains can cause intestinal infections, some of which are foodborne. Mechanisms of pathogenicity between the different strains of

E coli vary; toxin production, ability to adhere and invade, and other ill-understood processes are important in the production of different disease syndromes.

In the UK, verocytotoxin producing *E coli* (VTEC) infections are uncommon but they may be of increasing importance. The numbers of reported infections for 1987 and 1988 were 20 and 23, respectively; during 1989 more than 100 infections were reported. The disease (a haemorrhagic colitis) may be severe and haemolytic uraemic syndrome is a serious complication, especially in children. The reservoir in the UK is not known; the sources of outbreaks have not been clearly demonstrated, although in outbreaks in the USA, beef and milk have been implicated.[10]

Enteropathogenic *E coli* (EPEC), which cause disease in infants, may be transmitted via contaminated infant feeds or directly from infant to infant. They are of decreasing importance in the UK (Fig 3.1); about 1300 faecal isolates in children under 3 years old were reported in 1989.

Enterotoxigenic *E coli* (ETEC) cause disease in infants in countries where hygiene is poor and also in travellers to such countries. Man is the main source of infection, which is transmitted by contaminated food and water. The disease is uncommon in the UK.

Infection with enteroinvasive *E coli* (EIEC) is also uncommon in the UK, although occasional outbreaks occur in institutions. Infection is usually foodborne but person-to-person transmission may also occur.

OTHER ORGANISMS

Yersinia enterocolitica. Infections due to this organism are increasing in England and Wales; in 1989 about 580 cases of *Yersinia* infection were reported to CDSC (excluding *Y pseudotuberculosis*). The diarrhoeal illness usually affects children under 5 years of age. Reactive arthritis is a complication, especially in individuals of HLA type B27. The organism is found in wild and domestic animals and the infection seems to be spread by contaminated food and milk.[11]

Bacillus spp. The incidence of foodborne disease due to *Bacillus* species is low in England and Wales; in 1986, 1987, and 1988, the numbers of reported cases were 65, 137, and 418, respectively, most of which were due to *B cereus*. There are two forms of disease due to *B cereus*; the incidence of the diarrhoeal type (which is usually associated with meat products, vegetables, and puddings) is lower than that of the emetic type (in which the food most often implicated is cooked rice). The incidence of reported foodborne illness due to other *Bacillus* spp (*B subtilis*, *B licheniformis*, and *B pumilus*) in the UK has been low but constant over the years;[12] from 1975 to 1988, there were 103 episodes, involving 612 patients, reported to the PHLS Food Hygiene Laboratory.

Clostridium perfringens. This widely distributed anaerobic organism forms heat-resistant spores and is found in the gut of man and animals and in the environment. Food-poisoning results from eating contaminated food in which the organism has been allowed to multiply, as in cooked meat held at room temperature. There is recent evidence that person-to-person spread may occur among elderly people in hospital. From 1986 to 1988, the number of reported cases increased by 46% to 1312. Prevention is by good kitchen hygiene.

Viral foodborne disease

Hepatitis A virus is spread by the faecal/oral route and via contaminated food and water. Infection is endemic in countries where hygienic conditions are low. In England about 20-30% of young adults have serological evidence of infection. Outbreaks have especially been associated with shellfish, but other foods may act as the vehicle of transmission.[13] The increase in notifications during the past two years may be due to improved ascertainment.

Much less is known about hepatitis E virus, which also seems to be spread by the faecal/oral route, and there is at present little evidence of its occurrence in the UK.

Infections with astroviruses, caliciviruses, and "small round structured viruses" are endemic in the UK; outbreaks are due to person-to-person spread and some are food borne. In foodborne outbreaks, the food vehicle may be primarily contaminated, as with oysters and other shellfish, or may be contaminated by a human excretor. Among shellfish, oysters are particularly important because they are almost always eaten raw. That viruses do not multiply in food suggests that the infectious dose is small. Diseases due to these viruses are probably more seriously under-reported than are other intestinal infections because of the relative infrequency and difficulty of virological investigations; it is likely that they constitute a common cause of human disease. However, only about 370 cases were reported in 1989.

Prevention consists of good kitchen and catering hygiene to prevent faecal contamination of food and safer procedures for cultivation and treatment of shellfish;[14] cultivation of oysters in clean water is especially important.

Foodborne intoxications

STAPHYLOCOCCUS AUREUS

Although this organism predominantly causes skin infections, it is also associated with toxin-mediated illnesses, one of which is staphylococcal food-poisoning. The route of contamination is from an individual who harbours the appropriate strain to food, where the organism multiplies and produces toxin. The onset of the illness, in which vomiting predominates, occurs rapidly after ingestion of the food. *S aureus* is not at present an important cause of toxin-mediated food-poisoning in the UK, about 100 cases being reported in 1989. Prevention is by hygienic food preparation and exclusion of foodhandlers with staphylococcal skin disease.

CLOSTRIDIUM BOTULINUM

Botulism is rare in the UK. The disease is due to ingestion of contaminated food containing the botulinum neurotoxin. An unusual outbreak in 1989 (27 cases, 1 death) was associated with hazelnut yoghurt. In the previous ten years only 1 isolated case had been reported. Prevention is by the effective heat processing of canned foods and control of the preparation of other commercially and home-prepared preserved and bottled foods. Surveillance of this dangerous disease is important so that risks which may be associated with changing food processing techniques can be rapidly identified.[15]

Conclusion

Foodborne infection has substantial impact on public health. Continued surveillance and research is needed to learn more about such infection and how it can best be prevented; efforts must be maintained to ensure that established prevention methods are applied to all parts of the food chain.

It is not possible to acknowledge individually all those whose work has contributed to this review. However, I thank the CDSC, particularly the gastrointestinal section, and, within the Central Public Health Laboratory, the Food Hygiene Laboratory and the Division of Enteric Pathogens. I am also indebted to Dr J. W. G. Smith for his advice.

References

1. Threlfall EJ, Frost JA. The identification, typing and fingerprinting of *Salmonella*; laboratory aspects and epidemiological application. *J Appl Bacteriol* 1990; **68**: 5-16.
2. Memorandum of evidence to the Agriculture Committee Inquiry on Salmonella in Eggs. *PHLS Microbiol Dig* 1989; **6**: 1-9.
3. Agriculture Committee First Report. Salmonella in eggs—A progress report. House of Commons Paper 33. London: HM Stationery Office, 1989.
4. The microbiological safety of food. Part I. Report of the Committee on the Microbiological Safety of Food. HM Stationery Office, 1990.
5. The microbiological safety of food. Part II. Report of the Committee on the Microbiological Safety of Food. HM Stationery Office, 1991.
6. Skirrow MB. Campylobacter perspectives. *PHLS Microbiol Dig* 1989; **6**: 113-17.
7. Report of the World Health Organisation Informal Working Group on Food-borne Listeriosis. 15-19 February, 1988. Geneva: WHO, 1988. WHO/EHE/FOS/88-5.
8. Social Services Committee. Sixth Report. Food-poisoning; listeria and listeriosis. House of Commons Paper 257. London: HM Stationery Office, 1989.
9. Gilbert RJ, Hall SM, Taylor AG. Listeriosis update. *PHLS Microbiol Dig* 1989; **6**: 33-37.
10. Smith HR, Rowe B, Gross RJ, Fry NH, Scotland SM. Haemorrhagic colitis and verocytotoxin-producing *Escherichia coli* in England and Wales. *Lancet* 1987; **i**: 1062-65.
11. Anon. Yersinosis today. *Lancet* 1984; **i**: 84-85.
12. Kramer JM, Gilbert RJ. *Bacillus cereus* and other *Bacillus* species. In: Doyle MP, ed. Foodborne bacterial pathogens. New York: Marcel Dekker, 1989: 21-69.
13. O'Mahony MC, Gooch CD, Smyth DA, Thrussell AJ, Bartlett CLR, Noah ND. Epidemic hepatitis A from cockles. *Lancet* 1983; **i**: 518-20.
14. Appleton H. Small round viruses; classification and role in food-borne infections. In: Bock J, Whelar J, eds. Novel diarrhoea viruses. (Ciba Foundation Symposium 128) Chichester: Wiley, 1987: 108-25.
15. O'Mahony M, Mitchell E, Gilbert RJ, et al. An outbreak of foodborne botulism associated with hazelnut yoghurt. *Epidemiol Infect* 1990; **104**: 389-95.

4
Veterinary sources of foodborne illness

A. M. Johnston

Because microorganisms are widely distributed in animals and in foods of animal origin, control of food-borne infection is a formidable undertaking. The presence of many zoonotic diseases is often unsuspected or unrecognised in animals. On farms in the UK, especially the more intensively operated units, herd health schemes are run in cooperation with the veterinary practitioner and the State Veterinary Service where monitoring of disease is routinely carried out and active preventive medicine practised. Similar schemes are run in other developed countries. Factors which perpetuate problems include methods of husbandry (including the type of feedstuffs used), transport of live animals, and slaughter and meat processing practices (fig 4.1). Whichever system of farming is considered, it must be accepted that, however healthy our animals seem to be, pathogenic organisms are taken up from the environment, carried in the intestinal tract, and excreted in the faeces (fig 4.2). The aim, however, is to produce healthy animals, within the limits of humane rearing systems, with reasonable return on investment.

Animal enteric pathogens

The diseases of animals which affect the safety of food are predominantly those that cause enteric disorders. Soiling of the animal's coat with faeces will increase the possibility of subsequent contamination of food of animal origin. It is important that animals which produce milk or which will be slaughtered for food are kept as clean as possible.

Salmonella infection in animals produces an enteritis that affects the small and large intestines; many serovars, generally host-specific, have been identified. *Salmonella dublin* and *S typhimurium* are the most common serovars in ruminants. *S dublin* has a pronounced tendency to persist in cattle, which become intermittent or constant excretors or latent carriers; infection persists in the lymph nodes or tonsils, which can break down so that the

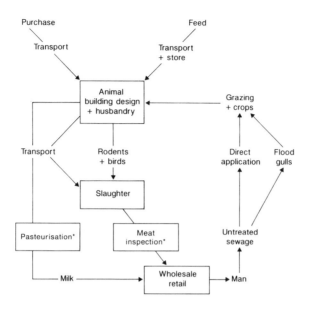

Figure 4.1 Veterinary aspects of control of foodborne illness.
*control stage

animals become excretors under stress. *S dublin* is a more common cause of abortion in sheep than *S abortus ovis*. Disease due to *S typhimurium* is sporadic and likely to subside after initial exposure, only recurring if the source of infection reappears. This does not, of course, mean that the disease cannot persist in a flock or herd for long periods. In sheep and cattle at grass, clinical salmonellosis can appear when the animals are concentrated in small areas such as lambing pens, holding pens on the farm, or in auction markets, especially when associated with transport. The situation in goats is similar to that in other ruminant species. *S cholerae-suis* infection in pigs has been especially well investigated following large outbreaks in UK pig units in the mid-1960s with high morbidity and mortality rates. Much of this infection was due to swill or garbage feeding which is no longer common practice. *S typhimurium* has also been found in pig units in which it presents as a necrotic enterocolitis, but, in abattoir surveys, other serovars have been found in the mesenteric lymph nodes and caecal contents.[1]

In poultry, salmonellae other than *S pullorum*, *S gallinarum*, *S enteritidis*, and *S typhimurium*, rarely cause clinical disease; the birds are carriers but show no overt features of disease. Clinical disease is seen in the young bird with substantial faecal excretion of organisms in recovered birds at time of slaughter as broilers—about seven weeks old.

Campylobacters can be isolated from the faeces of all animals, often without signs of clinical disease. In England and New Zealand, there is a seasonal prevalence of campylobacter infection in dairy cattle. There is some evidence of cross-infectivity between cattle and sheep of at least some

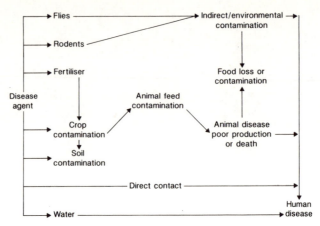

Figure 4.2   Sources of infection

strains.[2] *Campylobacter coli* and *C jejuni* have been implicated in enteritis, particularly in lambs in which morbidity rates have ranged from 20 to 75%, with mortality of about 3%. *C jejuni* causes abortion in sheep. Commonly, when campylobacters have been isolated from animals with diarrhoea, other pathogens (eg, *Cryptosporidium*, viruses, and other enteropathogenic bacteria) have been found.

Listeria species do not cause enteric disease but are found in the intestinal tract of animals and birds.[3] Herbivores probably ingest *Listeria* while eating grass, especially when it has been conserved in the form of silage; contamination is inevitable when machinery uplifts soil with the grass. There is a significantly higher prevalence of *L monocytogenes* in big-bale silage[4] because in this process more soil is taken up than in traditional forage harvesting for clamp fermentation. Since this product is more usually fed to sheep and goats, they are at greater risk of listerial infection. Septicaemic listeriosis, with or without meningitis, is most common in monogastric animals and young ruminants, whereas the meningeal cephalic form of the disease is more common in adult ruminants. In cattle, listeriosis tends to occur as single cases and sporadic abortions. Outbreaks in sheep and goats have been recorded soon after the start of silage feeding, with organisms subsequently being found in the faeces and milk.

All animal species are affected by *Escherichia coli*, the small intestine being the primary site of infection. The most important *E coli* are the enterotoxigenic K99 strains, which typically cause a distinct acute illness, known as watery scour, in young calves and piglets. Although enteropathogenic and enteroinvasive *E coli* are also found in animals, the enterohaemorrhagic *E coli* (EHEC) are of particular zoonotic importance. EHEC O157:H7 (which is verocytotoxigenic) is increasingly recognised in human disease and may be food borne, but other verocytotoxigenic *E coli* serotypes are more important in animals.[5] They produce diarrhoea in cattle and pigs (gut oedema disease). EHEC O157:H7 can certainly be found in the faeces of healthy cattle.

Yersinia spp are part of the normal gut flora of animals. *Y enterocolitica* can cause an acute enteritis with septicaemia and mesenteric lymphadenitis in sheep, goats, and pigs. *Y pseudotuberculosis* can lead to sporadic abortion in cattle and sheep and occasional cases of goat mastitis; it may produce enteric lesions and diarrhoea but more commonly causes multiple abscesses in the liver and the spleen.

Clostridium perfringens produces very acute disease with high mortality. Animal faeces are an important source of the organism, although the types of *Cl perfringens* associated with disease in animals rarely cause foodborne disease in man.

Cryptosporidium is an important enteric parasite in animals. This protozoan parasitises the gut of young livestock, which leads to severe diarrhoea and sometimes death. Some 30% of dairy herds in the UK are known to carry *Cryptosporidium*; from the age of about seven days, calves shed oocysts for about ten days but fewer than half of them have signs of disease unless other enteropathogens are involved. Subsequent re-infection of adults is possible from the young stock, with recrudescence after the stress of transport and lairaging (ie, animals kept at auction markets or abattoirs).

Giardia intestinalis is increasingly being found in the faeces of animals; little is known of its prevalence or of the disease it may cause in animals other than a chronic or intermittent diarrhoea, weight loss, and failure to thrive. *Giardia* has not been detected in pigs.[6]

Viral enteric diseases, especially those due to rotavirus and coronavirus, are important because they are responsible for a substantial number of neonatal diarrhoeas in animals, often with concurrent infections with Enterobacteriaceae.

Pathogens that affect the udder

Pathogens that affect the udder are important because they may get into the milk. Many infective agents have been implicated in mastitis. With the exception of tuberculosis, in which spread may be haematogenous, infection of the mammary gland is via the teat canal. Mastitis may be peracute or acute, with high mortality and morbidity; subacute, in which there is some change in the udder with clots and large numbers of leucocytes in the milk; chronic after an acute previous infection, with obvious changes to the udder and a variable extent of change in the milk; and subclinical, in which no change in the udder or milk can be seen.

Organisms which commonly cause mastitis in cattle and which are important with respect to foodborne illness include *Staphylococcus aureus* and *E coli*, with the latter becoming increasingly important in housed cattle. Other organisms relevant to foodborne illness that have been occasionally recorded as a cause of mastitis are *Salmonella*, *C jejuni*, and *L monocytogenes*. In the dairy cow, large numbers of *L monocytogenes* (25 000–30 000/ml) have been recorded in mastitic milk.[7] In surveys of raw milk from farm bulk tanks in north-east Scotland the incidence ranges from 2·8% of samples in the summer to 1% in October.[8] Wood and colleagues[9] described an *S enteritidis* infection involving a quarter of a cow's udder which continued to shed organisms into the milk; there was no faecal shedding of the organism during the seven months before slaughter.

Staph aureus is a common cause of mastitis in sheep and goats. Isolates from chronically infected goats have sometimes proved to be potentially enterotoxigenic and sheeps milk commonly contains toxin-producing *Staph aureus* (unpublished). *E coli* is increasingly found in mastitic milk; *Yersinia*, *Campylobacter*, and *Salmonella* are also occasionally found.

In addition to excretion of organisms in the milk, there is a risk during the milking process of contamination by faecal organisms. For example, *L monocytogenes* is regularly isolated from raw milk. Davidson and co-workers[10] found that 1·2% of samples from dairies and farms in Manitoba, Canada, were contaminated with *C jejuni* and 2·7% with *Y enterocolitica*. McEwen and colleagues[11] found that 4·7% of truckloads of bulk milk in Ontario, Canada, were contaminated with salmonellae (13 serovars), with a contamination at farm level of about 2·8%.

Feed

Animal feed and water can be a major risk to the animal population if they contain pathogenic organisms or toxins. Irrespective of how carefully animal feed is prepared, there may be subsequent contamination by wild animals and birds during transport and feeding. Completely closed feeding systems with silos and conveyor belts in intensive units reduce this risk compared with free-range systems, which provide free access for birds and vermin. Water has also been identified as a source of pathogens for livestock, particularly campylobacters and *Cryptosporidium*.[12,13] In the UK, use of animal protein derived from ruminants in feed for cattle and other ruminants has been banned after the association of bovine spongiform encephalopathy with feed.

Mycotoxins are present in pre-harvest and post-harvest food, particularly cereals, intended for animals. Mycotoxins such as aflatoxin B_1 and M_1 may then pass into the milk. Specific controls have been introduced to monitor known sources, such as groundnut and cotton seed cakes,[14] and maximum permitted levels for mycotoxins have been set.

Transport

The most severe stress for animals is associated with transport, especially when young calves (up to a week old) are transported and then exposed for sale in auction markets or taken to the abattoir as "bobby" (very young) calves. Stress of transport of poultry, which will increase the proportion of salmonella excretors, is reduced by the witholding of feed before transport. Correct transport in well-designed vehicles substantially reduces contamination of animals destined for slaughter; vehicles and crates should be cleaned and disinfected between loads. Movement of animals within and between countries brings additional problems of spread of disease. When diseases (important both to human and animal health) are present in the place of origin, animals must be put into quarantine or not introduced into a susceptible population of animals. Currently, harmonisation rules are being drawn up in the European Community so that there will be no compromise of the animal health status of any one member country.

Control

It must be accepted that food of animal origin is likely to contain organisms that are pathogenic for man, and continued efforts must be made to eradicate or reduce them to the minimum possible. Additionally, we must take into account that many organisms in the environment are pathogenic for the animals that we use for food. Although little can be done to remove these pathogens, it is possible to reduce further contamination of the environment by correct disposal of animal and human waste (Fig 4.2). Irrespective of the type of farm or species of animal, a reasonable return on a farmer's financial investment will only result from the use of good methods of husbandry and the maintenance of a high standard of herd/flock health.[15] Disease or poor herd or flock performance in animals used for food is ultimately expensive for the farming industry. Therefore, it is in the industry's interest as well as in the interest of public health to produce healthy animals. There should also be effective liaison between human and veterinary surveillance organisations so that the presence of potential human pathogens in animals used for food and the occurrence of foodborne disease in man can be monitored.

References

1. Tay SCK, Robinson RA, Pullen MM. Salmonella in the mesenteric lymph nodes and cecal contents of slaughtered sows. *J Food Protect* 1989; **52**: 202–03.
2. Meanger JD, Marshall RB. Seasonal prevalence of thermophilic *Campylobacter* infections in dairy cattle and a study of infection in sheep. *New Zealand Vet J* 1989; **37**: 18–20.
3. Skovgaard N, Morgen CA. Detection of *Listeria* spp in faeces from animals in feeds and raw foods of animal origin. *Int J Food Microbiol* 1988; **4**: 229–42.
4. Fenlon DR. Wild birds and silage as reservoirs of *Listeria* in the agricultural environment. *J Appl Bacteriol* 1985; **59**: 537–43.
5. Dorn CR, Scotland SM, Smith HR, Willshaw GA, Rowe B. Properties of verocytotoxin-producing *Escherichia coli* of human and animal origin belonging to serotypes other than O157:H7. *Epidemiol Infect* 1989; **103**: 83–95.
6. Kirkpatrick CE. Giardiasis in large animals. *Compendium Continuing Educ* 1989; **11**: 80–84.
7. Gitter M, Bradley R, Blampied PH. *Listeria monocytogenes* infection in bovine mastitis. *Vet Rec* 1980; **107**: 390–93.
8. Fenlon DR, Wilson J. The incidence of *Listeria monocytogenes* in raw milk from farm bulk tanks in North-East Scotland. *J Appl Bacteriol* 1989; **66**: 191–96.
9. Wood J, Chalmers G, Fenton R, Pritchard J, Shoonderwoerd M. *Salmonella enteritidis* from the udder of a cow. *Can Vet J* 1989; **30**: 833.
10. Davidson RJ, Sprung DW, Park CE, Rayman MK. Occurrence of *Listeria monocytogenes*, *Campylobacter* spp and *Yersinia enterocolitica* in Manitoba raw milk. *Can Inst Food Sci Technol J* 1989; **22**: 70–74.
11. McEwen SA, Martin SW, Clarge RC, Tamblyn E. A prevalence study of *Salmonella* in raw milk in Ontario, 1986-87. *J Food Protect* 1988; **51**: 963–65.
12. Reilly WJ. Human and animal salmonellosis in Scotland associated with environmental contaminants. *Vet Rec* 1981; **108**: 553–55.

13. Jones F, Watkins C. The water cycle as a source of pathogens. *J Appl Bacteriol* 1985; (suppl): 275–365.
14. Jewers K. Mycotoxins in food: the application of survey and quality control. *J R Soc Health* 1982; **3**: 114–18.
15. Johnston AM. Animal hygiene and husbandry in relation to food hygiene. *Outlook Agricult* 1990; **19**: 73–77.

5
Sources of infection: food

Diane Roberts

In the more developed countries there are some foods that can be regarded as "safe", having been made safe by the application of well-controlled decontamination processes, such as pasteurisation and sterilisation. Such foods include milk, ice cream, whole egg mix in bulk, and canned foods. Drinking and food-manufacturing water supply is controlled by filtration and chlorination. Other foods, such as bread, flour, jams, honey, pickles, fruits, and fats are regarded as safe because their composition, processing, or both provide conditions under which bacteria cannot multiply. The various properties of foods, such as pH, water activity (a_w), and salt or sugar content, can give an indication of whether microorganisms will grow. For example, most pathogenic bacteria will not grow in foods with a pH of less than 4·5, a low moisture content (a_w <0·86), or a high salt or sugar concentration. Many of these indices are inter-reactive and can be exploited in the preservation of foods.

However, we are constantly being encouraged to eat fresh rather than preserved foods, and it is this wide range of foods that may introduce into our kitchens many of the organisms that can lead to gastrointestinal illness if these items are not handled and stored correctly. There are three main routes by which microorganisms reach our food—namely, raw foodstuffs and ingredients, the foodhandler, and the environment.

Raw food and ingredients

MEATS

Foods of animal origin are the primary source of many of the bacteria responsible for foodborne infections and intoxications (Table 5.1). Organisms found in the live animal can be carried through to raw meats after slaughter, may persist through further processing, and ultimately may appear in the final retail product if insufficient attention is paid to hygiene and temperature control. Because there is mass rearing for food, a large proportion of animals reach the slaughterhouse excreting organisms such as *Salmonella* and *Campylobacter* in addition to the pathogens that form part of their normal faecal flora—namely, *Clostridium perfringens, Escherichia coli, Yersinia enterocolitica,* and *Listeria monocytogenes.* Continuous line processing, as for example in poultry processing establishments, increases spread from carcass to carcass so that many contaminated carcasses are distributed for retail or manufacturing purposes.

The rapid growth of the broiler industry has led to a more readily available and cheaper source of meat but also has increased infection in the birds and thus contamination of carcasses. 60–80% of retail chickens in the UK are contaminated with *Salmonella*[1] and reports from other countries indicate levels which range from 5 to 73%.[2] Up to 100% of birds may contain *Campylobacter*[3,4] and 60% may also harbour *L monocytogenes.*[5] Adequate heat treatment, which ensures centre temperatures of at least 70°C, will eliminate these organisms but sporing organisms such as *Cl perfringens* will still survive.

Red meats may also be contaminated but the prevalence of *Salmonella* and *Campylobacter* is probably lower than in poultry.[6] Comminution of meat will spread organisms throughout the product, so minced-meat items are a greater hazard.

EGGS

It has long been recognised that the shell of eggs may become contaminated with *Salmonella* from chicken faeces during laying and may be transferred to the bacteria-free contents during unhygienic breaking out. With the recent findings in the USA and the UK that *S enteritidis*-infected flocks lay a small proportion of eggs that have already become infected while in the hen's ovary,[1,7] there is a more serious problem with this food. Until the infection has been eradicated from laying flocks, raw egg must be regarded as a potentially hazardous food. Other European countries

Table 5.1   Principle food sources of the common food-poisoning organisms

Agent	Food source
Salmonella	Raw meat and poultry, raw milk, eggs
Clostridium perfringens	Meats, poultry, dried foods, herbs, spices, vegetables
Staphylococcus aureus	Cold foods (much handled during preparation), dairy products, especially if prepared from raw milk
Bacillus cereus and other *Bacillus* spp	Cereals, dried foods, dairy products, meat and meat products, herbs, spices, vegetables
Escherichia coli	Many raw foods
Vibrio parahaemolyticus	Raw and cooked fish, shellfish, and other seafoods
Yersinia enterocolitica	Raw meat and poultry, meat products, milk and milk products, vegetables
Campylobacter jejuni	Raw poultry, meat, raw or inadequately heat-treated milk, untreated water
Listeria monocytogenes	Meat, poultry, dairy products, vegetables, shellfish
Viruses*	Raw shellfish, cold foods prepared by infected foodhandlers

*For example, small round structured viruses, parvovirus, hepatitis virus.

have also had increasing reports of egg-associated *S enteritidis* infection—in particular, Spain.[1] The UK Government's health warning against the use of raw eggs in foods which will not receive further heat treatment should be heeded.

DAIRY PRODUCTS

Milk that has been treated (by pasteurisation, sterilisation, or ultra heat treatment) is a safe product, but raw milk from a healthy animal, which when freshly drawn contains very few bacteria, may acquire various pathogens from the milking animal or from the dairy environment and equipment. Outbreaks of milkborne infection, mainly of salmonellosis and campylobacteriosis, associated with the consumption of raw cows milk are still reported in England and Wales[8] and North America.[9] Raw milk is only rarely implicated in other European countries. The introduction of legislation in Scotland in 1983 which prohibits the sale of raw milk led to a large fall in the number of milkborne incidents. Occasional incidents of illness have been attributed to incorrectly pasteurised milk in some countries.[9-12] Products prepared from untreated milk are more likely to be contaminated than are those prepared from milk that has been heat treated. *Staphylococcus aureus, Bacillus cereus, Y enterocolitica*, and *L monocytogenes* may also be found in raw milk. All but *B cereus* will be removed by heat treatment; this organism, which produces heat-resistant spores, is a common dairy spoilage organism. However, incidents of *B cereus* intoxication from milk are uncommon, possibly because, owing to spoilage by the organism or other spoilage bacteria, the product becomes organoleptically unacceptable before it reaches the toxic stage.

Many of the methods of further acidification and fermentation of milk into other products will remove or inhibit most enteric pathogens, but a few organisms may survive. There are also further opportunities for the re-introduction of organisms by the addition of other untreated ingredients or from the production environment. Hard cheeses, yoghurt, and butter can be regarded as safe because of low pH and/or lack of moisture, but care must be taken with mould-ripened soft cheeses; these cheeses have a higher pH and thus permit the growth of *L monocytogenes*. 5–15% of soft ripened cheeses on sale in many countries in Europe—eg, the UK, the Netherlands, Germany, and Switzerland—may contain this organism[4,13,14] (usually in low numbers), irrespective of the heat treatment of the original milk.[13] The presence of the organism is most probably due to recontamination during the cheese making process or during handling at the distribution and retail stage.

FISH AND SHELLFISH

Fish and shellfish may become contaminated either from the environment from which they are harvested or from the environment during further processing. If such creatures are taken from water that has been polluted with sewage, they may contain many faecal pathogens (Table 5.1). *Vibrio parahaemolyticus*, a marine microorganism, is a common contaminant of raw fish and other seafoods, especially those from the Far East. It can be eliminated by heat treatment but poor production hygiene may lead to recontamination.

Shellfish are filter feeders and concentrate, in their bodies, organisms from the water in which they are breeding; microorganisms such as *Salmonella*, *E coli*, *V parahaemolyticus*, clostridia, and viruses have been detected. Thus, consumption of raw shellfish—eg, oysters—or shellfish that have been inadequately heat treated can lead to bacterial and viral infections.[15] These shellfish usually go through a depuration process in clean water whereby many organisms are washed out of the gills and bodies. This process is effective for bacteria but less so for viruses.

FRUITS, VEGETABLES, AND CEREALS

In the raw state these foods may be contaminated with any of the organisms present in the soil in which they are grown. Fruits that grow high above the ground are less likely to be contaminated than those that are in direct contact with the soil. The cleanliness of irrigation water also plays an important part in the extent of contamination of these products and in the pathogens that may be found. In countries where polluted water is used for irrigation and animal and human excreta for fertilisation, there is a risk of contamination with enteric bacterial pathogens, such as *Salmonella* (including *S typhi*), *Shigella*, and *V cholerae*, as well as viruses and parasites.[16] This is not a problem in the UK and developed countries where these practices are not common; in these countries most of the potentially harmful contaminating organisms are *Clostridium* spp, *Bacillus* spp, and *L monocytogenes*. For fruits and vegetables that are eaten without heat treatment, the extent of contamination can be reduced if they are washed and rinsed with a sanitising agent. Cooking will eliminate all but the sporing organisms.

DRIED FOODS

The predominant flora of most dried foods are organisms of the *Clostridium* and *Bacillus* groups, the spores of which survive the dehydration process. With some methods of drying, in which the temperature is not bactericidal or in which there are opportunities for recontamination, other organisms may persist—eg, *Salmonella* and *E coli*.[17] These foods are safe when dehydrated but as soon as they are rehydrated they must be treated as fresh foods. Since herbs and spices frequently carry heavy loads of sporing organisms,[18] it is important to add such ingredients at the beginning of the food processing stage (eg, before heat treatment).

READY-TO-EAT FOODS

In addition to our basic food items we are seeing a wide range of prepared, ready-to-eat foods retailed in either the frozen or chilled state. Many of these foods will have received some form of heat treatment but they are unlikely to be sterile. Frozen foods will remain safe while frozen but chilled foods require greater care with respect to shelf-life temperature of storage. The organisms found in these foods are those that have survived the heating process—eg, *Clostridium* spp and *Bacillus* spp—and those that have gained entry during the further manipulation of the product. The latter includes a wide range of organisms but those of concern are *Y enterocolitica* and *L monocytogenes* because they can grow at refrigeration temperatures. Numerous studies have found *L monocytogenes* in chilled foods,[19] including cooked chicken, complete meals, cooked meats, pâté,[20] and salads. Thus, greater care must be taken with this group of foods with respect to hygiene of production, temperature control during storage and display, and shelf life of the product.

The foodhandler and food environment

Organisms may also be transferred to food by the foodhandler either directly or by cross-contamination through the use of hands, surfaces, utensils, and equipment which have not been adequately cleaned and disinfected between the preparation of different foods. The factors which contributed to outbreaks of food-poisoning in England and Wales between 1970 and 1982 are shown in Table 5.2.[21] The infected foodhandler appears low on the list. Only in relation to *S aureus* food poisoning does the foodhandler have an important role. *S aureus* is frequently found in the nose and on the skin and can be readily transferred to foods by handling; subsequent storage of food at unsuitable temperatures will allow the organisms to multiply and produce their toxins. Thus, foods that are much handled during preparation and not reheated before final consumption are at greatest risk. Foodhandlers who continue to work with active symptoms of gastroenteritis are a hazard in the food preparation area because there is an increased risk of faecal organisms reaching food. Many people who constantly handle raw foods, particularly those of animal origin, often become symptom-free excretors and may carry the organisms on their hands. Provided they have good hand hygiene and formed stools, such individuals are not a major risk in food preparation.[22]

To prevent spread of organisms from man and the food environment, separate surfaces, equipment, and personnel should be used to deal with raw and cooked foods; there should be regular hand washing and good cleaning schedules that are regularly enforced. In large-scale food production, separate well-defined areas should be allocated to the preparation of raw and processed foods. Removal of food debris, which can attract insects and vermin, and thorough cleaning of all food preparation equipment (including the dismantling of items such as mixers and slicers so that all surfaces and crevices can be reached) followed by a final disinfection procedure and drying are essential to keep recontamination to a minimum. It is virtually impossible to eradicate all organisms from the food preparation environment but efforts must be made to keep levels as low as possible and to prevent outgrowth.

Every person involved in the preparation of food must be made aware that very few of the items that enter the kitchen are sterile. Most foods will be contaminated to a greater or lesser extent according to the amount of heat processing and other manipulation. If the food is consumed immediately, the extent of contamination is usually low—insufficient to cause illness. Most of the factors which contribute towards making a relatively harmless

Table 5.2   Factors that contributed to outbreaks of food-poisoning in England and Wales, 1970–82

Factors	% of outbreaks* in which factors recorded
Preparation too far in advance	57
Storage at ambient temperature	38
Inadequate cooling	32
Inadequate reheating	26
Contaminated processed food	17
Undercooking	15
Contaminated canned food	7
Inadequate thawing	6
Cross-contamination	6
Raw food consumed	6
Improper warm holding	5
Infected foodhandlers	4
Use of left overs	4
Extra large quantities prepared	3

*1479 outbreaks studied

food harmful can be grouped as either attributable to poor temperature control or to cross- contamination (Table 5.2). Thus, although improvement in agricultural procedures can lead to the reduction in the presence of some organisms in our foods (eg, *Salmonella* and *Campylobacter*), good hygienic food preparation and education of those involved in the preparation, processing, and service of food are the final lines of defence in the prevention of most types of foodborne illness.

References

1. Public Health Laboratory Service. Memorandum of evidence to the Agriculture Committee Inquiry on Salmonella in Eggs. *PHLS Microbiol Dig* 1989; **6:** 1–9.
2. D'Aoust J-Y. Salmonella. In: Doyle MP, ed. Foodborne bacterial pathogens. New York: Marcel Dekker, 1989: 327–445.
3. Hood AM, Pearson AD, Shahamat M. The extent of surface contamination of retailed chickens with *Campylobacter jejuni* serogroups. *Epidemiol Infect* 1988; **100:** 17–25.
4. Lammerding AM, Garcia MM, Mann ED, et al. Prevalence of *Salmonella* and thermophilic *Campylobacter* in fresh pork, beef, veal and poultry in Canada. *J Food Protect* 1988; **51:** 47–52.
5. Pini PN, Gilbert RJ. The occurrence in the UK of *Listeria* species in raw chickens and soft cheeses. *Int J Food Microbiol* 1988; **6:** 317–26.
6. Bolton FJ, Dawkins HC, Hutchinson DN. Biotypes and serogroups of thermophilic campylobacters isolated from ovine, bovine, and porcine offal and other red meats. In: Pearson AD, Skirrow MB, Lior H, Rowe B, eds. Campylobacter III. London: Public Health Laboratory Service, 1985: 106.
7. Humphrey TJ, Baskerville A, Mawer S, Rowe B, Hopper S. *Salmonella enteritidis* phage type 4 from the contents of intact eggs: a study involving naturally infected hens. *Epidemiol Infect* 1989; **103:** 415–23.
8. Barratt NJ. Milkborne diseases in England and Wales in the 1980s. *J Soc Dairy Technol* 1989; **42:** 4–6.
9. Sharp JCM. Infections associated with milk and dairy products in Europe and North America, 1980–85. *Bull WHO* 1987; **65:** 397–406.
10. Rampling A, Taylor CED, Warren RE. Safety of pasteurised milk. *Lancet* 1987; ii: 1209.

11. Jones PH, Willis AT, Robinson DA, Skirrow MB, Josephs DS. Campylobacter enteritis associated with the consumption of free school milk. *J Hyg (Camb)* 1981; **87**: 155–62.
12. Rampling A. The microbiology of milk and milk products. In: Linton AH, Dick HM, eds. Topley and Wilson's principles of bacteriology, virology and immunity. 8th ed. Vol 1. London: Edward Arnold, 1990: 265-89.
13. Greenwood M, Roberts D, Burden P. The occurrence of *Listeria* spp in milk and dairy products: a national survey in England and Wales. *Int J Food Microbiol* (in press).
14. Marth EH, Ryser ET. Occurrence of *Listeria* in foods: milk and dairy foods. In: Miller AJ, Smith JL, Somkuti GA, eds. Foodborne listeriosis. Amsterdam: Elsevier, 1990: 151–64.
15. Public Health Laboratory Service Working Party on Viral Gastroenteritis. Foodborne viral gastroenteritis. *PHLS Microbiol Dig* 1988; **5**: 69–75.
16. Geldreich EE, Bordner RH. Fecal contamination of fruits and vegetables during cultivation and processing for market. A review. *J Milk Food Technol* 1971; **34**: 184-95.
17. Rowe B, Begg NT, Hutchinson DN, et al. *Salmonella ealing* infections associated with consumption of infant dried milk. *Lancet* 1987; ii: 900-03.
18. Roberts D, Watson GN, Gilbert RJ. Contamination of food plants and plant products with bacteria of public health significance. In: Rhodes-Roberts M, Skinner FA, eds. Bacteria and Plants. Society for Applied Bacteriology symposium series No 10. 1982: 169-95.
19. Gilbert RJ, Hall SM, Taylor AG. Listeriosis update. *PHLS Microbiol Dig* 1989; **6**: 33-37.
20. Morris IJ, Ribeiro CD. *Listeria monocytogenes* and pâté. *Lancet* 1989; ii: 1285-86.
21. Roberts D. Factors contributing to outbreaks of foodborne infection and intoxication in England and Wales 1970–1982. Proceedings of the 2nd World Congress Foodborne Infections and Intoxications. Berlin: Institute of Veterinary Medicine, 1986; 157–59.
22. Cruickshank JG, Humphrey TJ. The carrier food handler and non-typhoid salmonellosis. *Epidemiol Infect* 1987; **98**: 223-30.

6
US food legislation

P. Thompson, P. A. Salsbury, C. Adams, and
D. L. Archer

The food available in the USA is abundant, wholesome, and diverse; it is among the safest in the world. These qualities can be attributed not only to considerable advances in science, medicine, and food technology, but also to the US laws, which govern the safety and economics of domestic and imported food products and thus contribute effectively to the prevention of foodborne illness.

Contemporary American food safety is a result of a long and often laborious evolutionary process that began in the 18th century and reached its first milestone in 1906 with the passage of the first pure food and drug act—the Federal Food and Drugs Act—and the Federal Meat Inspection Act in 1906/1907. During the past 85 years, these laws (designed primarily to deal with adulteration and economic fraud) have, with each modification, gradually emphasised the increasing concern of the American people about the effect of food on their health. These laws are administered by the Food and Drug Administration (FDA), an agency within the Public Health Service, the US Department of Health and Human Services, and the US Department of Agriculture (USDA) Food Safety and Inspection Service (FSIS) (Table 6.1). FDA traces its regulatory efforts in food protection against biological hazards to the original 1906 Food and Drugs Act, which was designed partly to protect food from contamination with harmful organisms. With the passage of this Act and the first Meat Inspection Act, the US Congress acknowledged that the wholesomeness and safety of the nation's food supply and the honesty of its labelling were legitimate concerns of the US Government.

The reporting of foodborne illness in the USA began in the 1920s when the Public Health Service prepared annual summaries of milkborne disease outbreaks reported by the states; in 1938, reports of waterborne and foodborne outbreaks were added.[1] The Centers for Disease Control, an agency of the Public Health Service, assumed responsibility for the publication of reports of foodborne illness in 1961.

In 1930, the Perishable Agricultural Commodities Act was passed by Congress. Currently administered by USDA's Agricultural Marketing Service, this Act makes compulsory the regulation of merchants, dealers, and brokers in the handling and marketing of perishable fresh fruits and vegetables in interstate and foreign commerce. 32 years after the first pure food and drug act, the Federal Food, Drug, and Cosmetic Act of 1938 (also referred to as the FD&C Act or the FDCA) was passed: it is currently administered by the FDA. This statute is the cornerstone of federal food safety policy, and it substantially revised the authority of the federal government to protect the public against adulterated and misbranded food products. For example, it provided the authority for food plant inspections. This Act remains the primary food safety law for FDA-regulated food products.

Adulteration is the term used in the FDCA for most safety-related non-labelling violations of the Act. Safety-oriented adulteration cannot be remedied by relabelling the adulterated goods.[2] Section 342 of the FDCA states that food is adulterated "if it contains any poisonous or deleterious substances which may render it injurious to health" (ie, pathogenic bacteria or their toxins). The same section states that a food is adulterated "if it consists in whole or in part of any filthy, putrid or decomposed substance or if it is otherwise unfit for food". Food may thus be adulterated under this section of the Act if it contains insects, animal excreta, or any substance deemed to be filth. Another part of Section 342 states that food "if it has been prepared, packed or held under insanitary conditions whereby it *may have become contaminated* with filth or rendered injurious to health" is also regarded as adulterated (our emphasis).

The 1944 Public Health Service Act (Section 243) provides for federal/state cooperative assistance in the prevention of interstate transmission of disease. Section 264 of this Act lays down regulations to prevent the introduction, transmission, or spread of communicable disease among the states. For purposes of carrying out and enforcing such regulations,

Table 6.1 US food legislation, codes, and regulations

Legislation	Code/Regulations
Federal Food and Drugs Act (1906)	. .*
Food, Drug, and Cosmetic Act (1938)	21 USC; Section 301 et seq
Perishable Agricultural Commodities Act (1930)	7 USC; Section 499 et seq
Public Health Service Act (1944)	42 USC; Section 243, 264
Poultry Products Inspection Act (1957)	21 USC; Section 451 et seq
Wholesome Poultry Products Act (1968)	21 USC; Section 451 et seq
Federal Meat Inspection Act (1906/1907)	21 USC; Section 601 et seq
Wholesome Meat Act (1967)	21 USC; Section 601 et seq
Food and Drug Administration Regulations (general and food related)	21 CFR; Parts 1–199
Food Safety Inspection Service Regulations	9 CFR; Parts 301 et seq and 9 CFR; Parts 581 et seq

*Superseded by the Food, Drug, and Cosmetic Act (1938).
USC = United States Code; CFR = Code of Federal Regulations.

the Act authorises inspection, fumigation, disinfection, sanitation, pest extermination, and destruction of animals or articles found to be sources of disease to human beings. Penalties also are authorised for failure to comply with these regulations. The Poultry Products Inspection Act, passed in 1957 because of greater US consumption of poultry and poultry products, ordered mandatory inspection of all poultry sold in interstate or foreign commerce. In 1968, the Wholesome Poultry Products Act extended USDA jurisdiction over poultry and poultry products to include poultry products sold within designated states, called for establishment of federal-state cooperative programmes of inspection, and established authority to withdraw or deny the grant of inspection. USDA jurisdiction over meat and meat products was likewise extended by the Wholesome Meat Act, which was passed in 1967.

FDA and FSIS programmes

FDA and FSIS programmes involve surveillance, enforcement, and prevention of foodborne safety problems; they also cover adulteration by rodents, bird and animal filth, and insect infestation.

FDA is responsible for ensuring the safety and wholesomeness of all foods sold in interstate commerce, apart from red meat and poultry, and has jurisdiction over products containing generally less than 2% cooked or generally less than 3% raw red meat or poultry, and other meat and poultry products not subject to the Federal Meat Inspection Act and the Poultry Products Inspection Act. FDA's routine sanitation inspections are the cornerstone of its food safety checks. The agency inspects domestic food plants, imported food products at US ports of entry, and feed mills that make medicated feeds for food-producing animals. It monitors recalls of unsafe or contaminated foods and can request court seizure of adulterated foods. FDA also provides guidance on good manufacturing practices (GMPs) for the food industry. Its GMP regulations (which deal mainly with sanitation) are supplemented by additional programmes, such as the hazard analysis critical control point (HACCP) system, which is a voluntary programme to avoid contamination in food processing, and provide guidelines, standards, and better methods to recognise conditions that violate the law (there is a mandatory programme for low-acid canned foods). In addition to routine sanitation inspections, microbiological hazards and sanitation problems are prevented through industry and consumer education, together with research designed to identify new hazards and to develop methods for their control.

FSIS is responsible for the inspection of meat and poultry products sold in interstate and international commerce for safety, wholesomeness, and label accuracy. The service ensures a sanitary environment in slaughter and processing plants and oversees the monitoring of all relevant stages of animal slaughter and meat and poultry processing procedures.[3,4] Since its inception, the meat inspection programme has made compulsory the examination both of individual animals before slaughter and of carcasses and vital organs for signs of disease. These examinations are made at the slaughterhouse at the time of slaughter. FSIS regulations also contain facility, equipment, and sanitary inspection requirements that are applicable to slaughter and processing operations. For these reasons, meat and poultry inspections have developed as "continuous" in-plant inspection programmes that operate on the manufacturers' premises. No meat or poultry manufacturer may operate without the provision of USDA-FSIS inspection. By contrast, regulations under the FDCA were, and still are, primarily concerned with the identification and exclusion of poisonous and deleterious substances from food.[5]

The USDA has powers to condemn foods, stop processing operations, obtain records, and approve all labels used on meat or poultry products.[5] Research is sponsored by USDA to improve food safety, support inspection operations, and improve food quality and availability. FSIS can recall unsafe or suspect products after they have reached the grocery shelf. FSIS also sponsors consumer education programmes to inform the public about proper care and handling of meat and poultry products.[3,4]

Microbiological standards

In 1988, USDA, in conjunction with FDA, the National Marine Fisheries Service of the US Department of Commerce, and the US Department of the Army, established the National Advisory Committee on Microbiological Criteria for Foods (NACMCF). The goals of the NACMCF are (a) to recommend, for specific commodities and products, HACCP procedures that can be used by food manufacturers and/or inspection operators to prevent or reduce contamination during the manufacture and distribution of their products, and (b) to develop and recommend microbiological criteria for foods when appropriate and applicable.

The most prominent microbiological standards formulated and enforced by federal regulatory agencies are those for milk, water, shellfish, and egg products. There are other standards for chemical contaminants, including animal drug residues, aflatoxin, scombrotoxin, and paralytic shellfish toxin in specific foods. At the federal level, there are no formal microbiological quality standards for other foods. Despite this lack of standards, existing law provides authority to remove from the marketplace products that are a threat to health. Therefore, "implied" microbiological standards do exist.[6] Generally, the fact that a certain type of food may be subject to other federal statutes does not mean that the requirements of the FDCA, the Federal Meat Inspection Act, the Poultry Products Inspection Act, or the Agricultural Marketing Act do not apply. In most cases, the statutes are not mutually exclusive; indeed, in many instances, they are jointly supportive. For example, the integrity of the various US standards for grades of fruits and vegetables by USDA is supported by FDCA prohibition against misbranding and adulteration. This, however, does not preclude FDA from relinquishing its jurisdiction, either wholly or partly, in instances where other federal agencies exercise more direct regulatory control over the specific article of food. This understanding between the agencies eliminates duplicative regulation and possible confusion in enforcement.

All 50 states and other federal agencies have less direct, but still important, responsibilities for helping to maintain the safety, including the microbiological safety, of the food supply. For example, products made from intrastate ingredients and sold entirely within a state are regulated by that state.

Conclusion

American food safety laws described here are fundamentally sound, but not perfect. Our basic laws are derived from early-to-mid-1900s science, modified by amendment. Each change was made in reaction to society's perception of a problem and resolved on the basis of then contemporary science. Assurances of food safety involve an inter-relation between science and law. Thus, advances in science can lead to the obsolescence of food laws that were based on limited scientific knowledge. Periodically, therefore, we examine those laws to be certain that these factors were taken into consideration so that regulation is realistic in view of the existing state of science.[7]

In the USA, there has been a resurgence of interest in microbiological hazards in its food supply because of several well-publicised foodborne disease outbreaks. This was strikingly illustrated in 1985 when *Salmonella* contamination in a dairy plant affected nearly 200 000 Americans. When accidental microbiological contamination happens on such a grand scale, the consequences are profound. Over the past 15 years, there has been an upward trend in the reporting of outbreaks and cases of foodborne disease due to microorganisms. Microbiological contamination of foods costs the USA at least $1 thousand million and possibly as much as $10 thousand million every year. Because of the rapid evolution and adaptability of microorganisms, and changes in methods of food production, processing, and storage, regulatory agencies need to develop effective microbiological control procedures.

Foodborne diseases are largely preventable. Knowledge of hygienic practices, of proper food processing techniques, and of epidemiology gives the public health community the tools necessary to prevent such diseases. However, that foodborne illness still occurs is not an indictment of the US food supply or its regulatory system, since there is, for example, an inherent risk associated with consumption of certain uncooked foods—eg, meats, poultry, eggs, fish, and shellfish. When adequately cooked, and unless recontamination occurs, the food is safe.[8] US federal and state regulatory agencies will continue to incorporate scientific advances and research to improve public health protection. The beneficiary is ultimately the American consumer who has access to an increasingly safer food supply.

References

1. Gilchrist A. Foodborne diseases and food safety. Wisconsin: American Medical Association, 1981: 2.
2. O'Reilly JT. Food and Drug Administration (Regulatory Manual Series). Colorado: Shepard's/McGraw Hill, 1979; Section 9–3.
3. National Research Council. Meat and poultry inspection: the scientific basis of the nation's program. Washington, DC: National Academy Press, 1985: 1.

4. National Research Council. Poultry inspection: the basis for a risk-assessment approach. Washington, DC: National Academy Press, 1987: 1, 13.
5. Olsson PC, Johnson DR. Meat and poultry inspection: wholesomeness, integrity and productivity. In: Food and Drug Law Institute, ed. 75th Anniversary Commemorative Volume of Food and Drug Law. Washington, DC: Food and Drug Law Institute, 1984: 220–41.
6. National Research Council. An evaluation of the role of microbiological criteria for foods and food ingredients. Washington, DC: National Academy of Sciencies, 1985: 48–52.
7. Food and Drug Administration. A brief history of U.S. food and cosmetic acts and regulations administered by FDA. Washington, DC: Center for Food Safety and Applied Nutrition, 1987: 22.
8. Archer DL, Young FE. Contemporary issues: diseases with a food vector. *Clin Microbiol Rev* 1988; **311:** 377–98.

7
UK food legislation

C. J. Ryder

British food is varied, wholesome, nutritious, of good quality, and is as safe as any in the world. This results from the continuing application of advances in science and technology, and from a body of legal requirements that has evolved over many years. In this review I shall discuss the provisions of food safety law, animal health law, and public health law that are most relevant to the prevention and control of foodborne illness, including the underlying principles, and I shall highlight current and foreseeable developments. The legislation that covers other problems of toxicity, such as lead in food, is not included, although the main statutory provisions are the same. These laws are variously the responsibility of the Food and Health Ministers, who act jointly on most aspects of food safety law.

Although the legislation described in this review provides the backbone of the system of control, various other means are used to secure appropriate standards of hygiene during food production. Central government, the enforcement authorities, industry, and other organisations produce various codes of practice, guidelines, and advisory booklets to improve awareness (both in the industry and among the general public) of potential risks associated with food and the ways to reduce them to a minimum. The sanction of prosecution underpins the system but is, effectively, used only when efforts to promote the safe and hygienic production of food have failed.

Food safety legislation

For at least the past 50 years the basic legislative approach has been to lay down in primary legislation (Acts of Parliament) a series of general requirements for food safety and consumer protection. There are also powers for Ministers to make secondary legislation (regulations) on more specific matters, in addition to enforcement powers for local authorities. The overall effect is to make those who produce, handle, and sell food responsible for ensuring that it is safe, fit, and of the requisite quality. Those who do not meet this responsibility can be prosecuted either for one of the general offences in the primary legislation or for an offence under the body of detailed Regulations that have been made under successive parent Acts.

PRIMARY PROVISIONS: THE FOOD SAFETY ACT 1990

The primary legislative provisions for Great Britain have recently been substantially strengthened and updated in the Food Safety Act 1990, which covers food from the moment of slaughter or harvest to the point of sale to the consumer. The provisions of the Act mostly came into force on Jan 1, 1991, and the general offences are now (i) rendering food by whatever means injurious to health with the intention that it should be sold for human consumption (Section 7); (ii) selling food (or supplying food in the course of business) that does not comply with food safety requirements—in particular, food that has been made injurious to health or is unfit for human consumption, or is so contaminated that it would be unreasonable to expect it to be used for human consumption (Section 8); (iii) selling food which is not of the nature, substance, or quality demanded by the purchaser (Section 14); and (iv) falsely or misleadingly labelling, advertising, or presenting food (Section 15).

The Act also provides powers which enable Ministers to make regulations on a wide range of subjects:

1 *Food safety and consumer protection.* Specifically, regulations may cover the composition or contamination of food or food sources, microbiological standards, food processes or treatments, hygienic conditions and practices (including training of foodhandlers), labelling, and such other provisions as may be necessary to secure compliance with food safety requirements (Section 16).
2 *Implementation of European Community (EC) obligations* under Section 17.
3 *The control of novel foods or of novel food sources, the prohibition of imports, and the power to make special provisions relating to milk* under Section 18.
4 *The registration by enforcement authorities of premises* used or proposed for use as food businesses and (where Ministers consider it necessary or expedient to do so in the context of food safety, public health, or consumer protection) for the licensing of food businesses and for the prohibition of the use of premises or processes except in accordance with the terms of licences (Section 19).

Government Ministers are also empowered to make emergency control orders on a type of food, food source, or contact material when they believe that there is an imminent risk of danger to health (Section 13). This power took effect as soon as the Act received Royal Assent on June 29, 1990. Additionally, the Food Safety Act 1990 provides local authorities with a series of new and much-strengthened enforcement powers: (i) enforcement officers may issue improvement notices under which proprietors of businesses must take specified steps within a given time to rectify hygiene problems (Section 10); (ii) prohibition orders may be imposed by the Courts when a defendant is convicted and when there is a risk of injury to health (Section 11); and (iii) when there is an imminent risk of injury to health, enforcement authorities can exercise emergency prohibition powers with immediate effect, although they must apply within 3 days for confirmation by a Magistrate's Court (Section 12).

Finally, the Act gives powers (in Section 40) for Ministers to issue statutory codes of practice to guide enforcement authorities towards an even and consistent standard of enforcement. These codes must be circulated for consultation before they are issued. The first codes, which came into force with the Act from January, 1991, will cover the new enforcement powers mentioned above; further codes will follow.

SECONDARY PROVISIONS

UK secondary legislation relevant to the microbiological safety of food lays down requirements on the hygienic handling of food during production and distribution, rather than setting maximum levels for the presence of microorganisms in food. This approach has been taken because for most microorganisms there is no expert consensus on minimum infective doses for all individuals, and also because the ability of microorganisms to multiply (or decrease) in food makes the imposition of limits at a certain point in the production chain an ineffective control on the level of microorganisms in food reaching the consumer. There has, however, lately been a move towards statutory microbiological standards, partly because this approach is favoured within the EC and partly because there is growing recognition that microbiological monitoring is one way to ensure that food processes are working properly and hygienically. The following are the main statutory instruments (SIs) that are relevant to food hygiene in England and Wales. Except where stated, similar provisions apply in Scotland and Northern Ireland.

The Imported Food Regulations 1984 (SI 1984 No 1981) broadly require that the importation of food for human consumption should take place under strict conditions. There are specific health certification requirements for meat and meat products. Enforcement of the regulations lies mainly with port health authorities.

The Food Hygiene (General) Regulations 1970 (SI 1970 No 1172) apply to fixed food premises and cover cleanliness and construction; the hygienic handling of food; the personal cleanliness of foodhandlers, and the notification of the local Consultant in Communicable Disease Control (CCDC) when handlers are found to have certain infections; the provision of a water supply, sanitary conveniences, and washing facilties; the proper disposal of waste material; and the temperatures at which certain foods should be kept on catering premises. Markets, stalls, and delivery vehicles are dealt with in a separate set of Regulations.

The Food Hygiene (Amendment) Regulations 1990 (SI 1990 No 1431) will, in a phased programme, introduce more stringent requirements on storage temperatures for a wide range of foods and extend the temperature requirements to cover the retail sector and delivery vehicles. All the foods covered will have to be kept at or below 8°C from April 1, 1991. Some of these foods will have to be kept at or below 5°C from April 1, 1993. There are various special dispensations—eg, for delivery vehicles of less than 7·5 tonnes and certain catering practices. These regulations were made in view of the latest knowledge of how *Listeria monocytogenes* and some other pathogenic bacteria can multiply at low temperatures.

The Poultry Meat (Hygiene) Regulations 1976 (SI 1976 No 1209) as amended implement EC requirements for intra-Community trade in fresh poultry meat, but apply also to meat that is intended for the domestic market. These regulations set detailed requirements for the production, cutting up, storage, and transportation of poultry meat. Slaughterhouses and cutting premises must (with certain exemptions) be licensed, and must meet detailed structural standards. Veterinary supervision is required, and antemortem and postmortem inspection standards are laid down.

The Fresh Meat Exports (Hygiene and Inspection) Regulations 1987 (SI 1987 No 2237) implement the EC Directive on intra-Community trade in fresh red meat; they lay down the structural and hygiene standards to be met by slaughterhouses, cutting premises, cold stores, and trans-shipment centres seeking to gain approval to export to EC member states. They also lay down meat inspection arrangements (both antemortem and postmortem) for slaughterhouses that have export approval. Such premises must be under veterinary supervision.

The Meat Inspection Regulations 1987 (SI 1987 No 2236) apply to non-export slaughterhouses, and required a postmortem, but not an antemortem, inspection. Ministers have, however, decided that antemortem inspection should be done in all slaughterhouses from Jan 1, 1991, and the Regulations have been amended to this effect (Meat Inspection [Amendment] Regulation 1990 [SI 1990 No 2495]). These regulations also set out detailed structural and hygiene standards for all slaughterhouses whether producing for export or for the domestic market alone.

The Liquid Egg (Pasteurisation) Regulations 1963 (SI 1963 No 1530) require liquid whole egg to be pasteurised when it is to be used in food that will be sold for human consumption, apart from when egg is broken out on the food manufacturer's premises and used within 24 hours. Proposals were issued in June, 1990, for replacement regulations that would extend the legislation to liquid yolk and albumen, introduce microbiological standards for egg products, ban the use of packing station mélange (broken eggs), and tighten control on the use of eggs by bakers and caterers.

The Milk and Dairies (General) Regulations 1959 (SI 1959 No 277) and the Milk (Special Designation) Regulations 1989 (SI 1989 No 2383). Extensive and detailed requirements have been developed over the years; initially they dealt with tuberculosis, but more recently have led to the recognition that the UK's milk production meets the highest microbiological standards set by the EC. The main requirements are as follows. The General Regulations require the registration of all dairy farms and all premises used as dairies. Registration is conditional upon the attainment of prescribed structural standards, the provision of safe and adequate water supplies, and the maintenance of hygienic equipment and milking routines. The Special Designation Regulations provide for the licensing of processing dairies to carry out pasteurisation, sterilisation, and ultra-heat treatment of milk. These Regulations specify the treatment processes as well as microbiological standards both for raw milk before treatment and for the heat-treated milk. They also set the conditions—recently reinforced with more stringent labelling and testing requirements—for the granting of licences in England and Wales to sell raw milk directly for human consumption. In Scotland, the sale of raw milk for direct human consumption is not permitted.

The Food Act 1984 (Section 35) prohibits the sale of milk from cows with acute mastitis, actinomycosis or suppuration of the udder, any infection of the udder or teats that is likely to transmit disease to man, any comatose condition, any septic condition of the uterus, anthrax, or foot-and-mouth disease. These provisions will be transferred to secondary legislation made under the Food Safety Act 1990.

Animal health legislation

The Zoonoses Order 1989 (SI 1989 No 285) re-enacts the provisions of an earlier Order by designating salmonellae and brucellae as risks to human health. This is a procedure provided for in the Animal Health Act 1981, so that some of this Act's powers can be used to deal with risks both to human health and to animal health. The 1989 Order also contained new provisions, which were the first stage in a set of linked measures dealing with human disease due to certain invasive salmonellae in poultry flocks.

The main effects of the measures are as follows. First, isolations of salmonella from an animal or bird, its carcass, products, or surroundings, or from feedingstuffs must be reported to the State Veterinary Service, which has been given extended powers to impose restrictions to control the spread of salmonella (Zoonoses Order 1989). Secondly, animal protein processors must, each day that they dispatch finished product from their premises, take a sample for testing at a laboratory approved by the Minister; and if this is positive for salmonella they must withhold the processed product from incorporation into animal feedingstuffs pending an official investigation. Processors must register with the Ministry of Agriculture, Fisheries and Food under the Processed Animal Protein Order 1989 (SI 1989 no 661). Finally, in relation to poultry, keepers of laying flocks, breeding flocks, or hatcheries must take samples and have them tested for salmonella according to a specified method. Laying flocks consisting of at least 100 birds, breeding flocks of at least 25 birds, and hatcheries with an incubating capacity of at least 1000 eggs must also be registered. These requirements are laid down in The Poultry Laying Flocks (Testing and Registration &c) Order 1989 (SI 1989 no 1964), and the Poultry Breeding Flocks and Hatcheries (Registration and Testing) Order 1989 (SI 1989 No 1963). A power to slaughter and pay compensation under the Zoonoses Order, 1989, was applied from March 1, 1989, to laying flocks in which *Salmonella enteritidis* or *S typhimurium* were confirmed, and was subsequently extended to breeding flocks producing birds for egg or for broiler production. During January, 1991, the automatic slaughter policy for laying flocks infected with *S typhimurium* was discontinued because of the extremely low number of outbreaks of food-poisoning associated with *S typhimurium* in eggs.

Public health legislation

The requirements for the notification and control of infectious diseases, including food-poisoning, are in the Public Health (Control of Disease) Act, 1984, and the Public Health (Infectious Diseases) Regulations, 1988. The implication in these titles that the legislation is recent is misleading. The 1984 Act largely consolidated earlier provisions, some dating back to the late 19th and early 20th century. The legislation is currently under review with a view to introducing new legislation better geared to modern needs.

These provisions apply notification requirements and control measures to specified infectious diseases. The entirety of the control powers in the 1984 Act are applicable only to the five diseases listed in Section 10—namely, cholera, plague, relapsing fever, smallpox, and typhus. For food-poisoning (made notifiable under Section 11) and a further 24 diseases made notifiable under the 1988 Regulations, the available control powers are more limited; those for food-poisoning are in Sections 11, 12, 18, and 20 of the Act and in Schedule 4 of the 1988 Regulations.

A doctor who diagnoses or suspects that his patient has a notifiable disease or food-poisoning must inform the proper officer appointed by the local authority (usually the CCDC). Since notification is required on suspicion, doctors ought to notify on clinical diagnosis and not wait for laboratory confirmation. The proper officer can then act promptly to identify the source of infection and prevent further cases.

Current and future developments

The Food Safety Act 1990 has considerably strengthened regulation-making and enforcement powers across the board, among other things extending the range of possibilities for action in relation to microbiological food safety. The Government is also constantly developing the secondary legislation—eg, the recent updating of the temperature controls in the Food Hygiene Regulations. The Government also intends to make early use of the powers of the new Food Safety Act to introduce for the first time regulations on the training of foodhandlers and on the registration of food premises. Registration will enable enforcement authorities to identify all permanent and commercial food businesses in their area, to rank them according to the risk they present, and to set priorities for their enforcement activities accordingly. The Government is also bringing the licensing powers of the Act into operation. Regulations were laid down on Dec 11, 1990, subjecting the application of food irradiation to strict licensing controls. The Government has also announced its intention of regulating production, storage, and distribution of hermetically-sealed products because of concern over the possibility of growth of *Clostridium botulinum*.

Future developments will also be crucially influenced by the report of the Committee on the Microbiological Safety of Food (Richmond report). The Committee made many detailed recommendations in the two parts of its report published on Feb 15, 1990, and Jan 14, 1991.[1,2] In particular, the Government is implementing the central recommendation for a national microbiological surveillance and assessment system underpinned by a new permanent committee structure. This system will lead to further developments in the system of control, based on the results of research and surveillance.

EUROPEAN COMMUNITY

Another very important influence shaping future legislation will be the EC, which has some measures already in place and a large number of proposals under consideration. The EC recognises that national governments should legislate for food safety, but insists that legislation should not create barriers to trade by requiring the same food to meet different requirements in different member states. This principle has been reinforced by a series of judgments in the European Court of Justice, which affirm that, unless clearly justified on grounds of public health or consumer protection, national measures that ban the importation of food lawfully produced and marketed in another member state are illegal under the Treaty of Rome even if such measures result from requirements that are applied indiscriminately to home and imported produce.

The activities of the EC have lately assumed a different character and a greater importance as progress is made towards completion of the internal market. Up to 1986, the Council of Ministers generally adopted legislation on food safety, animal health, or plant health unanimously; now it tends to act by qualified majority voting. The scope for variation in member states' requirements is rapidly diminishing as more and more areas become subject to harmonised Community control; thus, national legislation is precluded except if it is required or authorised by the EC legislation. Self-evidently, the existence of EC Commission proposals for Community legislation will in itself place considerable constraints on member states' scope for introducing controls unilaterally. Even an absence of Commission proposals on a given subject does not leave member states free to legislate since national measures must be notified in draft to the EC Commission (under Directive 83/189/EEC on technical standards, as extended to food by Directive 88/182/EEC). After notification, there must be a standstill of 3 months for the EC Commission and other member states to comment. Depending on the comments made, the standstill period can be extended to 12 months. The aim is to enable the Community to take its own measures, rather than individual member states acting unilaterally.

One important area where the EC is active in developing legislation is the hygiene of products of animal origin. A series of Directives, starting in 1964 with Directive 64/433/EEC which covered fresh meat, lays down requirements for the hygienic production of various products. They cover such matters as structural requirements for premises; hygiene requirements for processes, equipment, and staff; registration of premises; regular official inspections; and, sometimes, microbiological standards and temperature control. To date measures have been adopted on red meat, poultry meat, meat products, minced meat, egg products, and milk and milk products. Their implementation in UK law—either actual or proposed—is through the various statutory instruments briefly outlined above.

The EC Commission's Agriculture Directorate (DG VI) is further developing this group of measures in preparation for the single European market. Some of the original measures applied only to products traded between EC member states, and these now need to be extended to cover all EC output. DG VI has also made proposals (or has declared its intention of doing so) to cover the products of animal origin that are not yet the subject of specific measures—ie, game and rabbit meat, animal fats and by-products of rendering, fish products, live bivalve molluscs, and other products of animal origin for human consumption.

Current and proposed Community measures, which set out general requirements that are not specific to particular products, are also relevant. Directive 89/397/EEC on the Official Control of Foodstuffs requires the regular inspection of all stages of the production, manufacture, importation, processing, storage, transport, distribution, and trading of foodstuffs. The EC Commission's Internal Market Directorate (DG III) is planning a Directive on food hygiene, and has taken some informal soundings of public views. This Directive seems likely to result in a proposal for hygiene standards for the production and marketing of all foodstuffs. DG VI has already issued its proposal for a regulation (see above) on products of animal origin, which contains elements laying down general hygiene provisions. Care will be needed in negotiations to ensure that any eventual measures are compatible and complementary. DG VI may also propose zoonoses controls.

Conclusion

Many EC measures are still subject to negotiation, and so it remains to be seen what requirements will eventually result. However, amendment to existing UK legislation will probably be needed to the extent that Community provisions when adopted do not mirror our own. Decisions by the EC will be the major influence, alongside developing technical knowledge, on the future development of the UK legal provisions which bear on foodborne illness.

I thank the many colleagues in the Agriculture and Health Departments (too numerous to mention individually) who have helped me with comments on successive drafts of this article.

References

1. The microbiological safety of food. Part 1. Report of the Committee on the Microbiological Safety of Food. London: HM Stationery Office, 1990.
2. The microbiological safety of food. Part II. Report of the Committee on the Microbiological Safety of Food. London: HM Stationery Office, 1991.

8
Foodborne salmonellosis

A. C. Baird-Parker

Salmonellas continue to be one of the main causes of foodborne illness world wide; in many countries salmonellosis is the most frequently reported foodborne disease. In various western countries, including the UK, reported cases of salmonellosis have increased steadily during the 1970s and 1980s, with an especially sharp rise in some countries during the past five years due to certain phage types of *Salmonella enteritidis*. Thus, the socioeconomic cost of human and animal salmonellosis adds substantially to the cost of food production and health care, and causes much suffering and financial loss. There is international recognition of the need to find cost-effective solutions.[1]

Bacteriology

Salmonellas are gram-negative, motile, rod-shaped bacteria that can grow both aerobically and anaerobically between about 7 and 48°C (optimum 37°C), at pH 4–8, and at water activities above 0·93. They are readily killed by heat (eg, 71·7°C for 15 s) and acid (eg, 1·4% acetic acid at pH 4·0 within 72 h), and are resistant to both freezing and drying, particularly in the presence of proteins and other protectants.[2]

The organism is ubiquitous among domestic and wild warm-blooded animals and almost all serovars cause illness in man. Some types are adapted to certain animal species; these types are much more virulent for the species to which they are adapted than for other species. For example, the avian-adapted *S pullorum* and *S gallinarum* cause severe disease in poultry but very rarely infect man. *S enteritidis* phage type (PT) 4 causes an invasive infection of poultry that leads to septicaemia and subsequent chronic infection of various organs; when the ovary is infected, transmission of organisms to the contents of the egg can occur. Although highly virulent for poultry, *S enteritidis* PT4 does not have enhanced virulence for man.[3]

Although there are more than 2200 serovars of salmonella, fewer than

200 cause human illness in the UK in any one year. *S typhimurium* and *S enteritidis* currently account for three-quarters of reported cases; *S enteritidis* accounts for more than half the total number of cases reported. These 2 serovars are also the main cause of human infection in many other western countries. The serovars that lead to illness in man generally reflect those found in the environment, especially in farmed and wild animals and in birds. Further division of serovars is done by phage typing or plasmid typing and is essential for epidemiological studies and for the investigation of outbreaks.

Incidence

In 1989, there were more than 26 000 cases of salmonellosis in England and Wales (Public Health Laboratory Service [PHLS] Communicable Disease Surveillance Centre, unpublished). About 10% of salmonella infections are contracted overseas.[4] The current reported rate of infection in Great Britain is about 500 cases/million population; this is believed to be substantially lower than the true incidence rate, which is the subject of a pilot study set up by the Department of Health, based on general practitioner consultations in which clinical diagnosis is confirmed microbiologically.[4] Although the number of cases in England and Wales approximately doubled between 1985 and 1989, they seem to be levelling off; this is evidenced not only by a smaller increase in incidence in 1989 than in 1988, but also by a negligible overall increase in incidence (<1%) in 1990, although there was a substantial increase in infections due to *S enteritidis*, particularly PT4. The number of cases also increased in Scotland and Northern Ireland during the 1980s but the rises were smaller. There have been peaks in salmonellosis previously—eg, during the 1950s in England and Wales, and in the early 1980s in Scotland when there was a large increase due to *S typhimurium* contamination of raw milk.

The increase in salmonellosis in Great Britain since 1984 has been almost entirely due to *S enteritidis*. More than 60% of salmonellas from human cases typed by the PHLS Division of Enteric Pathogens between January and August, 1990, were *S enteritidis*, 85% of which were PT4. *S enteritidis* PT4 has been found in 16% of strains from broiler chickens on retail sale,[5] and in the contents of at least 1% of eggs from infected poultry flocks (battery and free range).[6,7] Clearly, *S enteritidis* PT4 has reached epidemic proportions in the UK, with an 8½-fold increase in cases from 1984 to 1989. The increase continued with a 25% increase in incidence in 1990.

Various European countries have also reported a higher incidence of *S enteritidis* PT4 infections; the organism has been isolated from human cases and food items from at least ten European countries and from countries as far apart as Argentina and Japan. In parts of the USA there have also been large increases in *S enteritidis* infections associated with shell eggs, but the types involved have been mainly PT8 and PT13.[8] The incidence of salmonella in countries with comparable reporting systems is similar: in 1987, the number of cases per million in West Germany was 640, in Norway 390, in the USA 200, and in the UK 360. In all countries the incidence of salmonellosis is highest during the summer. Enteric fever is now uncommon in northern Europe; most cases in the UK are imported from Asia and the Indian subcontinent.

Salmonellosis

ENTERITIS

The principal symptoms of infection in man are diarrhoea, abdominal pain, mild fever, chills, nausea, and vomiting; prostration, anorexia, headaches, and malaise may also occur. The incubation period is 5–72 h (occasionally up to 7 days). Symptoms usually occur between 12 and 36 h after infection and usually last from 2 to 5 days. Illness is generally more severe in very young and elderly people and death is more likely among these age groups. The infective dose in man is at least 1 million cells. However, there is strong epidemiological evidence that many fewer organisms—eg, 10–100 cells—can cause illness in very young children and elderly people, especially if the organism is contained in high-fat foods, such as cheese, chocolate, hamburgers, or salami.[9,10] Infection from water contaminated with low numbers has also been reported.

Although the acute stage of illness passes fairly rapidly, the carrier state can last for more than 3 months and sometimes there are complications. For instance, in patients with underlying disease, such as cancer, septicaemia is not uncommon and in healthy subjects there may be a wide range of sequelae, including pericarditis, neurological and neuromuscular diseases, reactive arthritis, ankylosing spondylitis, and osteomyelitis. Also, there may be damage to the mucous membrane of the small intestine and colon, which leads to malabsorption and nutrient loss; allergies and severe illness are more likely in malnourished individuals.[11,12]

Management of clinical salmonellosis is usually by fluid replacement therapy. Antibiotic treatment is likely to prolong the carrier state and is therefore not recommended in cases with no complications. Ciprofloxacin reduces the duration of diarrhoea and fever, and eliminates salmonella from stools; the sugar derivative lactulose is useful in some cases. The organism can usually be isolated from the vomitus and faeces, but not the blood, although some strains of *S dublin*, *S typhimurium*, and *S cholerae-suis* often cause septicaemia.

ENTERIC FEVERS

Enteric fevers are due to *S typhi*, *S paratyphi* A, *S paratyphi* B, and *S paratyphi* C. The principal features are malaise, headache, high and persistent fever, body aches, general weakness, and abdominal pains; nausea, vomiting, coughing, sweating, chills, and anorexia may also occur. The incubation period is 7–28 days (on average 14 days), depending on the infective dose.

Pyrexia can persist for several weeks when the patient may become delirious. Septicaemia generally occurs 10 days after onset; the organism can be detected in blood and urine samples, with positive stools often 3 weeks after infection. The carrier state, which is more common in female and elderly patients, may persist for several months and occasionally for many years.

The infective dose is thought to be low. There seem to be little data about host and vehicle effects but these are probably the same as those for other salmonellas. The gall bladder is the usual seat of infection in the carrier state, although other organs, such as the liver, may be involved. The organism is excreted intermittently in the urine and faeces. Diagnosis is usually by blood culture and the detection of the Vi antigen in blood samples.

Costs

Direct costs of salmonellosis include loss of earnings, costs of diagnosis and investigation of an outbreak, and costs of recall and destruction of the implicated food when a common food-source outbreak is identified. Indirect costs, such as compensation for pain and suffering of victims, should also be taken into account. In some countries legal fees may be a substantial part of the cost of an incident.[13]

The estimated annual cost of salmonellosis in the USA is about US$1·4 billion (about US$700/case).[14] In 1986 estimated costs in the UK for treatment, diagnosis, and investigations were £375/case.[15] In North America, costs per case range from about US$800 for incidents associated with restaurants, take-away meals, and food prepared in the home to about US$10 000 for cases associated with a food producer; costs for typhoid incidents were US$10 000 and US$350 000, respectively.[16]

The benefit of introducing measures that reduce or prevent the occurrence of salmonellosis often outweighs the costs. For example, based on data from a large outbreak of salmonellosis due to raw milk in Scotland in 1981, the estimated costs of purchasing and operating pasteurising equipment would be much lower than the annual cost of potential salmonella cases associated with raw milk. Likewise, several studies have demonstrated the positive cost/benefit of irradiation for treatment of poultry. For example, in a cost/benefit comparison of potential salmonella control measures in Canada,[1] the annual cost of irradiation of processed poultry was Can$18·5 million, whereas the benefit in terms of reduction in human salmonellosis was estimated to be Can$52·7 million. Roberts et al[17] calculated that, in an outbreak associated with the contamination of chocolate with *S napoli*, successful investigation yielded a 3·5-fold rate of return to the public sector.

Sources and transmission

Salmonellosis in animals and man is usually due to consumption of contaminated food or water; secondary spread may occur directly via the faecal-oral route—eg, among intensely farmed animals and among patients in hospitals and institutions. Salmonellas are zoonotic organisms carried in the intestinal tract and associated organs of most farm and wild animals, including mammals, birds, reptiles, amphibians, and arthropods. They usually cause little disease in their hosts but are excreted in large numbers in their faeces. The effluent from infected animals and man is an important source of contamination of the environment and the food chain.

The cycle of salmonella transmission is complicated (Fig 8.1). A farm animal may be infected from various sources, including feeding stuffs, birds (eg, gulls), bedding, flies, rodents, sewage, soil, and water. There may also be vertical transmission by host-adapted types. For example, *S enteritidis* PT4 can pass from breeding flocks to the production layers and broilers via transovarian infection of the egg contents or the egg shell during, or shortly after, laying. During transportation to slaughter, there may be cross-contamination of animals since stress during transportation will often increase the excretion of salmonella in the faeces. There are opportunities for contamination at many of the stages during the slaughter, dressing, and preparation of raw meats; these are very important sources of salmonella in the human food chain. Inedible materials and other parts of the animals not used for human food are sent to the rendering plant; a by-product of the process is the meat and bone meal used in compounded animal feeds. Salmonellas present in the raw materials before rendering will be destroyed by the rendering process but cross-contamination may occur. There are also further opportunities for contamination in the manufacture of animal feeds, although the process of "hot-pelleting" (extrusion of hot mash under pressure to form pellets) is effective for destroying any salmonella present in feed raw materials.

The prevalence of salmonella in foods is extremely variable. On average, about half of broiler carcasses are contaminated with salmonella, whereas contamination of beef and lamb carcasses is generally less than 1% (with higher contamination on pig carcasses) depending on the source.[18] The prevalence in raw red meats has generally decreased during the past twenty years. Cereals, salad, and other vegetables are occasionally contaminated with salmonella but the prevalence and contamination levels are much lower than in meat.

Prevention and control

The prevention and control of salmonellosis requires joint action not only by the agriculture and food sectors of the industry, but also by the consumer.

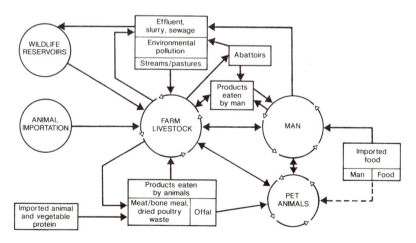

Figure 8.1   Cycle of salmonella transmission

The application of the hazard analysis critical control point (HACCP) system throughout the food chain to identify the best places (critical control points) to eliminate or control growth or contamination with salmonella is a pre-requisite for the effective and economic control of human and animal salmonellosis.[19,20]

Control starts on the farm and there are many opportunities to reduce to a minimum the carriage of salmonella in the animals we use for food. These include the use of salmonella-free stock and feed; good husbandry practices; protection of feed and water from contamination; the hygienic disposal of all waste; and better design of intensive rearing areas for easier cleaning before each use and to discourage pest infestation.

Wherever possible, outside areas should be protected from contamination; for example, the spraying of fresh slurry onto land where it could contaminate drinking water or grazing land, and the siting of refuse tips close to farms should be avoided. Vaccines are generally not effective for commercial immunisation of animals. The use of "competitive exclusion", in which flora from the intestinal tract of a mature animal is used to infect a very young animal (eg, a 1-day-old chick) to establish a protective flora, has been shown to be effective in some instances and is widely practised on poultry farms in Scandinavian countries.[8]

The UK Government has now introduced a series of measures to limit infection and spread of salmonellas in animals and in particular to limit the introduction of *S enteritidis* and *S typhimurium* into the human food chain.[4] Thus, any salmonellas isolated from designated live animals or carcasses, or from their products or surroundings must be reported (Zoonoses Order 1989). Domestically produced and imported animal proteins used for animal feeds have to be tested for salmonella (Processed Animal Protein Order 1989); similar requirements apply to the final feed. Additionally, laying birds in commercial flocks of 25 or more and breeding flocks used for both commercial layers and broilers must be registered and tested (Poultry Breeding Flocks and Hatcheries [Registration and Testing] Order 1989); there are powers to destroy any flock found contaminated with salmonella and regarded as a health hazard. Recently announced proposed European Community regulations extend these requirements to the rest of the community. Additionally, a series of UK Codes of Practice concerning the operation of rendering plants, transportation of animal products, manufacture of animal feeds, and procedures to be used in hatcheries and broiler houses have been drawn up jointly by industry and the Government.

Transportation of animals to abattoirs or markets should be done under conditions that minimise stress, and in vehicles and crates designed to be cleaned and disinfected between use. Similar requirements are needed for holding animals in stalls and lairages, where good stockmanship and the hygienic design and operation of holding areas are necessary to limit animal-to-animal spread of salmonellas.

Hygienic abattoir practices are extremely important with respect to the contamination of meat with salmonellas. Antemortem and postmortem inspection are important for the removal of diseased animals from the food chain, although the vast majority of animals carrying salmonella in their intestinal tracts do not show any visual signs of infection. Animals entering the slaughter line should be clean, and it is also important that the water in scalding tanks (to assist in the removal of hair or feathers) should be held at the highest temperature possible, where appropriate filtered, and replaced frequently to prevent the build-up of bacteria. The use of acids, alkalis, and other bactericidal agents in scalding tanks used for defeathering poultry has been claimed to reduce carcass infection but these generally have not been effective in commercial plants. Similar claims have been made for agents in water used for washing carcasses of beef animals; proper use of the singeing process to remove any hair remaining after the de-hairing process will also reduce surface contamination of pig carcasses.

Evisceration is probably the procedure that most commonly leads to the contamination of meat with salmonellas. The mechanical eviscerators in modern poultry plants, though designed to improve the efficiency and speed of evisceration, are also an effective means of spreading salmonella contamination. It is inevitable that occasionally the intestinal tract of a bird will rupture during its removal by the eviscerator; thus, other carcasses subsequently handled on the equipment will become contaminated, despite use of sound hygiene measures, including frequent and thorough rinsing of the machine and carcasses with potable water. Salmonellas adhere very well to the skin of chicken and are difficult to remove once they have become attached; they also are very resistant to disinfectants when thus attached. Because of this adhesion, even well-designed chillers remove hardly any salmonella, although when properly operated (with correct use of disinfectants, such as chlorine), chillers will minimise cross-contamination; air chilling also limits cross-contamination. There are similar difficulties, but to a lesser extent, with the manual systems used for the evisceration of red meat animals. Many techniques and devices are recommended to remove the intestinal tract intact; despite improvements made in recent years, there is inevitably a failure rate. The higher the line-speed in abattoirs, the greater is the possibility of rupture of the intestinal tract and difficulty in managing the slaughter and dressing process to reduce to a minimum microbial contamination of the carcass meat. The cooling of carcasses as quickly as possible after dressing (taking into account the requirements of quality) is important to avoid the growth of any salmonellas.

Despite attempts to reduce infection in live animals, and to apply good hygienic practices in the abattoir and during further processing, it is inevitable that some raw meats will be contaminated with salmonella. Irradiation can be used to eliminate the low numbers of salmonella that may be present on raw meats, although there are indications that this procedure may not presently be generally acceptable to consumers. The advantage of irradiation is that it can be applied to the final packed product and thus recontamination is substantially avoided. Since poor hygienic practices in the kitchen also contribute to salmonellosis, proper measures must be taken to limit recontamination of cooked food and cross-contamination to other foods, and to store foods that may be eaten without further cooking correctly to prevent growth of organisms. Education of the general public about basic food hygiene,[21,22] including education in schools, is essential to reduce the incidence of food-poisoning.

Although there are many ways to prevent the occurrence of salmonellas in the foods we eat, it must be accepted that raw foods, especially meats, will continue to be a source of salmonellas for many years. Thus, there is a need for proper care throughout the food chain to reduce the incidence of salmonellosis to the minimum level possible.

I thank Dr B. Rowe, Central Public Health Laboratory, UK, for kindly reading and commenting on this paper.

References

1. World Health Organisation. Salmonellosis control: The role of animal and product hygiene. *WHO Tech Rep Ser* 1988; no 774.
2. D'Aoust JY. Salmonella. In: Doyle MP, ed. Foodborne bacterial pathogens. New York: Marcel Dekker, 1989: 328–445.
3. Rampling A, Upson R, Ward LR, Anderson JR, Peters P, Rowe B. *Salmonella enteritidis* phage type 4 infection of broiler chickens: a hazard to public health. *Lancet* 1989; ii: 436–38.
4. The microbiological safety of food. Part 1. Report of the Committee on the Microbiological Safety of Food. London: HM Stationery Office, 1990: 116–25.
5. Memorandum of evidence to the Agriculture Committee Inquiry on Salmonella in Eggs. *PHLS Microbiol Dig* 1989; **6**: 1–9.
6. Mawer SL, Spain GE, Rowe B. *Salmonella enteritidis* phage type 4 and hens' eggs. *Lancet* 1989; i: 280–81.
7. Humphrey TJ, Cruickshank JG, Rowe B. *Salmonella enteritidis* phage type 4 and hens' eggs. *Lancet* 1989; i: 281.
8. World Health Organisation. Consultation on epidemiological emergency in poultry and egg salmonellosis. Geneva: WHO, 1989.
9. Blaser MJ, Newman LS. A review of human salmonellosis: 1 Infectious dose. *Rev Infect Dis* 1982; **4**; 1096–106.
10. D'Aoust JY. Infective dose of *Salmonella typhimurium* in cheddar cheese. *Am J Epidemiol* 1985; **122**: 717–20.
11. Isenberg HD. Pathogenicity and virulence: another view. *Clin Microbiol Rev* 1988; **1**: 40–53.
12. Archer DL. Diarrheal episodes and diarrheal diseases: acute disease with chronic implications. *J Food Protect* 1984; **47**: 322–28.
13. Todd ECD. Legal liability and its economic impact on the food industry. *J Food Protect* 1987; **50**: 1048–57.
14. Roberts T. Human illness costs of foodborne bacteria. *Am J Agricultural Econ* 1989; **71**: 468–74.
15. Sockett PN, Stanwell-Smith R. Cost analyses of the use of health care services by sporadic cases and family outbreaks of *Salmonella typhimurium* and campylobacter infections. Proceedings of the 2nd World Congress: Foodborne Infections and Intoxication, Berlin: WHO, 1986; **2**: 1036–39.
16. Todd ECD. Preliminary estimates of costs of foodborne disease in Canada and costs to reduce salmonellosis. *J Food Protect* 1989; **52**: 586–94.
17. Roberts JA, Sockett PN, Gell ON. Economic impact of a nationwide outbreak of salmonellosis: cost-benefit of early intervention. *Br Med J* 1989; **298**: 1227–30.
18. Mackey BM. The incidence of food poisoning bacteria in red meat and poultry in the United Kingdom. *Food Sci Technol Today* 1989; **3**: 246–49.
19. Simonsen B, Bryan FL, Christian JHB, Roberts TA, Tompkin RB, Silliker JH. Prevention and control of salmonellas through application of HACCP. *Int J Food Microbiol* 1987; **4**: 227–47.
20. International Commission on Microbiological Specifications for Food. Microorganisms in

Foods Vol 4: Application of the hazard analysis critical control point (HACCP) system to ensure microbiological safety and quality. Oxford: Blackwell Scientific, 1988.

21. World Health Organisation. Food safety: examples of health education materials. Geneva: WHO, 1989.

22. World Health Organisation. Food, environment and health; a guide for primary school teachers. Geneva: WHO, 1990.

9
Campylobacter

M. B. Skirrow

Campylobacters have escaped much of the recent publicity given to food-poisoning organisms in the UK. One reason may be unfamiliarity since campylobacter enteritis has been recognised for little more than a decade. Another reason is that campylobacters rarely cause death or spectacular outbreaks of food-poisoning, even though food is a common vehicle of infection. Nonetheless, they are the most frequently identified agents of acute infective diarrhoea in most developed countries, and thus they are a major burden on society with respect to health care costs, loss of productivity, and personal suffering.

Bacteriology

Campylobacters are only distantly related to most other medically important bacteria. A glance down a microscope at a living campylobacter culture bears this out, for one sees a melée of tiny darting rods, which, when slowed down or stationary for a moment, are seen to be spiral. This spiral shape facilitates colonisation of mucous membranes; it enables the organisms to "corkscrew" their way through mucus with a speed that cannot be matched by other bacteria. They are strictly microaerophilic and require special atmospheric conditions for cultivation in vitro. Special selective media are needed for isolation from clinical material such as faeces.

Campylobacter enteritis is caused by the two closely related species, *Campylobacter jejuni* and *C coli*, of which there are more than 100 serotypes. In most parts of the world, *C jejuni* is the predominant species (80–90% of infections), but the distinction is mainly of epidemiological interest since the disease each species produces is essentially the same. Recent work suggests that *"C upsaliensis"* is also enteropathogenic,[1] and occasionally other species, such as *C lari* and *C hyointestinalis*, are isolated from patients with diarrhoea.

Systemic campylobacteriosis is an uncommon subacute bacteraemic form of infection seen only in patients with immune deficiency or a predisposing abnormality such as a damaged heart valve. Although infections are believed to originate in the intestine, abdominal symptoms are usually absent. *C fetus* is typically associated with systemic campylobacteriosis, but *C jejuni* and other *Campylobacter* spp also cause this condition. The disorder is distinct from the transient bacteraemia that sometimes accompanies campylobacter enteritis in otherwise healthy people.

Campylobacter enteritis

Campylobacter enteritis is an acute enterocolitis with histological features indistinguishable from those of salmonella or shigella infection. Most campylobacter strains produce a cholera-like enterotoxin and one or more cytotoxins. Clinically, the disease cannot be distinguished from the other acute bacterial diarrhoeas, but the intensity and duration of abdominal pain is on average greater in campylobacter enteritis and there is often a non-specific febrile prodrome. The average incubation period is 3 days (range 1–7). Almost all patients recover within a week. Complications are uncommon, but patients with severe abdominal pain are sometimes admitted to hospital with suspected acute appendicitis, and reactive arthritis develops in about 1% of patients 1–2 weeks after onset of illness. Peripheral polyneuropathy (Guillain-Barré syndrome) is a less frequent but more serious complication.[4,5]

Treatment with antibiotics is seldom required in this self-limiting infection, but for severely affected patients erythromycin or ciprofloxacin are effective. Untreated patients excrete campylobacters in their faeces for 2–3 weeks (about half have negative stool cultures after 3 weeks), but transmissibility is low and it is only rarely necessary to restrict healthy foodhandlers who are excreting the organism. Chronic carriage is unknown in healthy people.

Incidence

In 1990, laboratories in England and Wales reported almost 35 000 campylobacter infections; this represents an annual incidence of about 87/100 000 (Public Health Laboratory Service, Communicable Disease Surveillance Centre, unpublished). Similar rates have been reported in Europe and North America where surveillance was of similar extent. Rates have remained fairly constant in the USA, but in England and Wales annual totals have quadrupled since 1979 and the upward trend shows no sign of lessening. However, much of this increase is due to increased testing—ie, we are merely detecting a higher proportion of the true total. We can only guess how big that total is, but conservative estimates place it at about ten times the reported number. This estimate is in line with a general practice survey, in which the projected annual rate of campylobacter enteritis was 1100/100 000.[6]

In developed countries a bimodal age distribution, with peaks of incidence in infants and young adults, is consistently found.[7,8] The secondary peak in young adults, which is especially pronounced in young men (M/F ratio 1·7/1 in one study[8]), is not usually seen in salmonella or shigella infection. The reasons for this pattern are unknown, but one suggestion is that it is due to the popularity of "fast" take-away foods among young people. The high incidence in adults points to a lack of herd immunity; by contrast, in developing countries, exposure to the organism is sufficiently high for immunity to be gained early in childhood.[7,9] In these countries, environmental contamination from domestic animals in the home is probably a major factor in transmission. People who live in rural areas, irrespective of country, have more infection than do urban dwellers. Because of the high prevalence of campylobacters in the developing world they are a common cause of travellers' diarrhoea.

SEASONAL TRENDS

In temperate regions there is an increased incidence in campylobacter infection in the summer. Records kept over 9 years in England and Wales show a strikingly consistent rise that begins in May, reaches a peak in July, and then falls steadily to base levels by December (Fig 9.1). This summer rise occurs 8 weeks ahead of the summer rise of salmonella infections. In some years there is a small secondary peak in the autumn, a feature also seen in the USA. However, the reasons for these patterns of infection are unknown.

Costs

Detailed costing of 53 patients with campylobacter enteritis in England gave an average figure of £273 per patient, or £587 if the intangible costs of pain and suffering are included.[10] Thus, the tangible cost alone of the 1990 total of 35 000 laboratory diagnosed cases amounts to nearly £10 million; and if

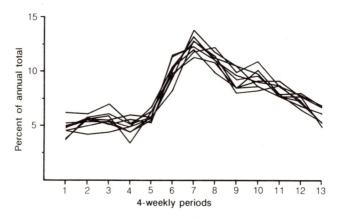

Figure 9.1 Composite graph to show seasonal pattern of campylobacter isolations. No of isolates reported to the PHLS Communicable Disease Surveillance Centre, UK, for each of the 9 years, 1981–89. Mean annual total 22 200.

the previously mentioned annual rate of 1100/100 000 in the general practice survey is representative of the whole country, the cost could be ten times as high.

Sources and transmission

C jejuni and *C coli* live in a wide range of animals. They are especially common in birds, an adaptation that is reflected in their high optimum growth temperature of 42–43°C. Wild birds are the principal reservoirs of infection for domestic animals and animals used for food, which in turn constitute the main source of infection for man. Occasionally, wild birds may be a more direct source of infection via water. Campylobacters are present in almost all surface water, which if distributed untreated can lead to major outbreaks of campylobacter enteritis.[7] In the UK, infection has also been acquired by the drinking of milk from doorstep delivered bottles that had their metal foil caps pecked by magpies (*Pica pica*) and jackdaws (*Corvus monedula*).[11,12] These birds behave in this curious way when they are feeding their young (May and June), and there are places where it accounts for a substantial proportion of seasonal infection.

The transmission of infection from domestic and food producing animals may be direct or indirect. Direct transmission is largely occupational and individuals who are regularly exposed develop immunity. In the home, infection from contact with puppies or kittens with campylobacter diarrhoea is well known but is probably infrequent. More important is indirect transmission via food. Campylobacters are introduced either on the flesh of animals or in milk. Like waterborne infection, milkborne infection tends to occur in outbreaks, whereas foodborne infection tends to be sporadic. Foodborne outbreaks are seldom reported and when they are they are usually small.

MILK

Campylobacters are commonly present in raw milk; they get there by faecal contamination at the time of milking, but occasionally cows excrete organisms into their milk from campylobacter mastitis. Either way, the distribution of raw or inadequately pasteurised milk has led to major outbreaks of campylobacter enteritis. In the UK alone, 13 such outbreaks were recorded in 3 years, the largest affecting some 3500 people.[13] The only certain prevention is to pasteurise or otherwise heat-treat all milk sold to the public.

ANIMAL MEAT

There are three ways by which people can become infected from animal flesh: (1) By the handling of the raw product. For example, infection has been reported in inexperienced workers in poultry processing factories, trainee chefs learning to dress uneviscerated chickens, and housewives preparing raw chickens. (2) By consumption of the raw or undercooked product. Raw meats, including fish and shellfish, have been implicated as sources of infection, but cooked meats are only a risk if there is obvious undercooking. Moreover, unlike salmonellas, campylobacters do not multiply in warm meat. Barbecue and fondue cooking methods carry an increased risk of infection. (3) By cross-contamination of "innocent" foods. The consumption of foods such as salads and bread that have become cross-contaminated from raw meats in the kitchen is probably the most frequent route of infection. The infectious dose can be as low as a few hundred bacteria.

Red meat and offal. Cattle, sheep, and pigs commonly carry campylobacters in the gastrointestinal tract and their carcasses regularly become contaminated at slaughter. Fortunately, air chilling greatly reduces numbers of campylobacters by means of surface drying. Thus, contamination rates of red meats at the point of retail sale are low: 1·4% found in one survey is typical,[14] but higher figures have been reported.[15] Offal (liver, kidney, heart) is more commonly contaminated (6–47%). Pigs, which typically carry *C coli* rather than *C jejuni*, are probably an important source of infection in countries where pork products are eaten salted or lightly cooked.

Poultry. Since birds are natural hosts of *C jejuni* and *C coli*, colonisation of poultry is almost inevitable. The problem is exacerbated by the ease with which contamination occurs during mass mechanised processing of poultry carcasses. Thus, campylobacters can be grown from most poultry sold in shops. In one study, fresh chickens had campylobacter counts of up to $1·5 \times 10^6$ per bird and uneviscerated chickens up to $2·4 \times 10^7$ per bird; frozen chickens had fewer organisms.[16] A case-control study in Seattle, USA, attributed 48% of all cases of campylobacter enteritis to the handling and consumption of chickens.[17] Similar studies need to be done elsewhere, but results are unlikely to be very different in Europe and other developed areas, where eating habits are similar to those in Seattle.

Control

The single most effective measure to control campylobacter enteritis would be to control infection in broiler chickens. Much could be done to reduce human infection by education of the public in good hygienic practice, and to a lesser extent by a reduction in contamination during mass processing, but these are palliative measures and the final solution must be to produce birds that are free of or only lightly colonised with campylobacters.

At present we do not know how this can be done because we do not know how broiler chickens become infected. Campylobacters are most probably introduced in growing flocks from extraneous sources since infection is not vertically transmitted via eggs. Recent work in the UK implicated broiler house water supplies and distribution systems as the main route by which infection is introduced and maintained.[18] The combined skills of microbiologists, veterinarians, and engineers working with the poultry industry are needed to confirm this and find a solution. This will be costly, but the outlay is likely to be small compared with the cost of allowing the disease to continue unchecked.

References

1. Goossens H, Vlaes L, De Boeck M, et al. Is *"Campylobacter upsaliensis"* an unrecognised cause of human diarrhoea? *Lancet* 1990; **335**: 584–86.
2. Skirrow MB. Campylobacter infections of man. In: Easmon CSF, Jeljaszewicz J, eds. Medical microbiology. Vol 4. London: Academic Press, 1984: 105–41.
3. Walker RI, Caldwell MB, Lee EC, Guerry P, Trust TJ, Ruiz-Palacios GM. Pathophysiology of *Campylobacter* enteritis. *Microbiol Rev* 1986; **50**: 81–94.
4. Kaldor J, Speed BR. Guillain-Barré syndrome and *Campylobacter jejuni*: a serological study. *Br Med J* 1984; **288**: 1867–70.
5. Winer JB, Hughes RAC, Anderson MJ, Jones DM, Kangro H, Watkins RPF. A prospective study of acute idiopathic neuropathy. II. Antecedent events. *J Neurol Neurosurg Psychiatry* 1988; **51**: 613–18.
6. Kendall EJC, Tanner EI. Campylobacter in general practice. *J Hyg (Camb)* 1982; **88**: 155–63.
7. Blaser MJ, Taylor DN, Feldman RA. Epidemiology of *Campylobacter jejuni* infections. *Epidemiol Rev* 1983; **5**: 157–76.
8. Skirrow MB. A demographic survey of campylobacter, salmonella and shigella infections in England. A Public Health Laboratory Service Survey. *Epidemiol Infect* 1987; **99**: 647–57.
9. Calva JJ, Ruiz-Palacios GM, Lopez-Vidal AB, et al. Cohort study of intestinal infection with campylobacter in Mexican children. *Lancet* 1988; i: 503–06.
10. Sockett PN, Pearson AD. Cost implications of human campylobacter infections. In: Kaijser B, Falsen E, eds. Campylobacter IV: Proceedings of the fourth international workshop on campylobacter infections. Göteborg: University of Göteborg, 1988: 261–64.
11. Hudson SJ, Sobo AO, Russel K, Lightfoot NF. Jackdaws as potential source of milk-borne *Campylobacter jejuni* infection. *Lancet* 1990; **335**: 1160.
12. Southern JP, Smith RMM, Palmer SR. Bird attack on milk bottles: possible mode of transmission of *Campylobacter jejuni* to man. *Lancet* 1990; **336**: 1425–27.
13. Robinson DA, Jones DM. Milk-borne campylobacter infection. *Br Med J* 1981; **282**: 1374–76.
14. Bolton FJ, Dawkins HC, Hutchinson DN. Biotypes and serotypes of thermophilic campylobacters isolated from cattle, sheep and pig offal and other red meats. *J Hyg (Camb)* 1985; **95**: 1–6.
15. Fricker CR, Park RWA. A two-year study of the distribution of 'thermophilic' campylobacters in human, environmental and food samples from the Reading area with particular reference to toxin production and heat-stable serotype. *J Appl Bacteriol* 1989; **66**: 477–90.
16. Hood AM, Pearson AD, Shahamat M. The extent of surface contamination of retailed chickens with *Campylobacter jejuni* serogroups. *Epidemiol Infect* 1988; **100**: 17–25.
17. Harris NV, Weiss NS, Nolan CM. The role of poultry and meats in the etiology of *Campylobacter jejuni/coli* enteritis. *Am J Public Health* 1986; **76**: 407–11.
18. Pearson AD, Colwell RR, Rollins DM, et al. Prevention of *C jejuni* transmission to broiler chicken by farm interventions. In: Kaijser B, Falsen E, eds. Campylobacter IV: proceedings of the fourth international workshop on campylobacter infections. Göteborg: University of Göteborg, 1988: 304–05.

10
Foodborne listeriosis

Dorothy Jones

Listeria monocytogenes has been recognised as an animal and human pathogen for more than 60 years.[1] Well known in veterinary practice, the bacterium was for many years regarded by clinicians as something of a curiosity. The recent increase in incidence of human listeriosis and the association of some of these cases with contaminated food has generated much public concern and drawn the attention of clinical and food microbiologists to the causative organism.

Bacteriology

Of the seven currently recognised species of the genus *Listeria*,[2,3] only *L monocytogenes* is an important human and animal pathogen. The other species are generally believed to be non-pathogenic, although *L seeligeri*, *L welshimeri*, and *L ivanovii* (a sheep pathogen) have been found rarely in human infection.[4] All listeriae are morphologically and physiologically very similar, but nucleic acid studies indicate that the genus consists of two closely related but distinct lines of descent[2] (Table 10.1).

Primary isolation is not always easy from normally sterile sites such as blood, and can be even more difficult from contaminated materials such as faeces or environmental samples.[4,5] However, once isolated, the bacteria grow well on the usual laboratory media. Listeriae are facultatively anaerobic; they grow better in air but anaerobic conditions favour the recovery of partly damaged cells. Carbohydrates are essential for growth. The optimum growth temperature is 30–35°C (neutral or slightly alkaline pH) with a wide range (0·5–45°C) but there is much strain variation at the higher temperatures. Despite recent controversy, *L monocytogenes* does not survive properly conducted, commercial pasteurisation.[6] The ability to grow at low temperatures has been exploited in the successful, but time consuming, cold enrichment technique (4°C) for isolation of listeriae.[5] This ability has also led to concern that foods stored at refrigeration temperatures,

Table 10.1 Differential characters of *Listeria* spp.

Species grouping	Species	Haemolysis* on horse blood	CAMP test†		Acid from‡				Reduction of nitrate to nitrite†
			Staphylococcus aureus	*Rhodococcus equi*	D-mannitol	L-rhamnose	D-xylose	α-methyl D-mannoside	
1	L monocytogenes	+	+	−	−	+	−	+	−
	L innocua	−	−	−	−	V	−	+	−
	L ivanovii	++	−	+	−	−	+	−	−
	L seeligeri	W	+	−	−	−	+	V	−
	L welshimeri	−	−	−	−	V	+	+	−
2	L grayi	−	−	−	+	−	−	++	−
	L murrayi	−	−	−	+	V	−	++	+

* + = moderate; + + = strong; W = weak.
† + = positive; − = negative.
‡ + = acid produced; − = acid not produced; V = variable reaction.

particularly those consumed without subsequent cooking, are an important source of human infection. Most strains grow at pH 9·6 but all are inhibited to some extent at pH below 5·5. Inhibition at low pH seems to be dependent on the nature of the acid; organic acids (acetic > lactic > citric) are more inhibitory than hydrochloric acid;[7] and this accounts for the low viability of transfers made from cultures used in fermentation studies and for the low levels or absence of listeriae in adequately fermented foods and silage. Listeriae grow in 10% sodium chloride, survive for up to a year in 16–20% sodium chloride, and survive long periods of drying and of freezing with subsequent thawing; these characteristics favour survival, if not growth, in the environment.

Listeriae are short (0·5×1–2 μm), regular, gram-positive, non-acidfast, non-sporing rods that occur singly or in short chains. However, palisade and Y-form arrangements of some cells can lead to initial confusion with *Corynebacterium*. Similarly, the occasional rod more than 10 μm in length resembles *Erysipelothrix*, and the coccoid forms, often seen in broth cultures or in smears from infected tissue, can be mistaken for streptococci. Confusion can also arise from the gram-negative appearance of some cells in direct smears from incompletely treated clinical cases and older cultures. Listeriae have a characteristic tumbling motility when cultured at 20–25°C but not at 35°C.[3,8] All pathogenic strains of *L monocytogenes* lyse a range of mammalian red blood cells.

There are at least 16 serovars of listeriae in the current scheme based on the serological grouping of 14 heat-stable somatic (O) antigens and four heat-labile flagellar (H) antigens.[9] This scheme does not include the serologically different species *L grayi* and *L murrayi*. With the exception of *L ivanovii* (of which all strains are serovar 5), there is no strict correlation between serovar and species; *L monocytogenes* and *L seeligeri* in particular share many antigens.[3,8] Therefore, serological methods alone are not useful for identification to species level. *L monocytogenes* can be divided into 13 serovars, of which 3 (1/2a, 1/2b, 4b) are responsible for most human cases;[4,8,10] serotyping is therefore of very limited epidemiological use. Bacteriophage lysotyping gives excellent discrimination between strains but its value is limited because fewer than 70% of strains of *L monocytogenes* are usually typable by available bacteriophages.[11] The recent application of multilocus enzyme electrophoresis (MEE) for the analysis of clinical and food isolates of *L monocytogenes* in both Europe and the USA[12,13] seems to be very promising for epidemiological studies. The use of DNA restriction enzyme analysis[14] and monoclonal antibodies[15] for typing *L monocytogenes* strains for epidemiological purposes are also being assessed.

Human listeriosis

Listeriosis is a rare illness with a high mortality rate.[4,5] Clinical features range from mild influenza-like illness to meningitis and meningo-encephalitis. The striking increased monocytosis noted in the first human case of listeriosis and frequently seen in some animal infections is rare in man. Illness is most likely to develop in pregnant women, and in very young, elderly, or immunocompromised individuals, but apparently healthy young adults may also be affected. In non-pregnant adults, men seem to be more susceptible.[4,8] According to McLauchlin[10] the incidence of listeriosis in the UK is highest in the autumn, whereas Gellin and Broome have not found a pronounced seasonal occurrence in the USA. The incubation period varies from 1 to 70 days.

Infection in pregnant women usually leads to a mild influenza-like illness; meningitis is extremely rare. Infection of the fetus can lead to abortion, stillbirth, or delivery of an acutely ill infant. Neonatal listeriosis is either early onset (within 2–3 days of birth) after infection in utero or late onset (>85 days) after infection during birth or in the neonatal period. Early-onset listeriosis is characterised by pneumonia, septicaemia, and disseminated abscesses with high mortality (about 40–50%). In late-onset cases, meningitis is the most prominent feature and the mortality rate is lower (about 25%).[4]

In non-pregnant immunosuppressed individuals, meningitis or bacteraemia/septicaemia are the most common clinical features. Although listeriosis is uncommon in AIDS patients, investigators have estimated that the risk of such patients getting listeriosis is several hundred times greater than that of the general population.[4] Listeriosis in apparently healthy individuals almost always occurs as meningitis. The world-wide estimated mortality rate in the more severe forms of the disease is about 30%—higher in the elderly and the immunocompromised population.[4]

Focal infections caused by *L monocytogenes* are very rare and are mainly seen in immunocompromised individuals; such infections are probably due to seeding during an initial bacteraemia. The bacterium can cause mild localised cutaneous lesions without systemic involvement in veterinarians, farmers, and poultry workers who attend or handle infected animals or birds. Conjunctivitis has also been reported in poultry workers and after accidental laboratory infection; such infections are usually mild and localised but sometimes lead to meningitis.[4,5]

The antibiotic treatment of choice is ampicillin, or ampicillin with an aminoglycoside; alternatives, for individuals allergic to beta-lactams, are tetracycline, erythromycin, or chloramphenicol (alone or in combination).[16] However, even with intense antibiotic therapy, prognosis in the most severe forms of listeriosis is poor. Antibiotic resistance is rare.[3,4] The recent detection of transferable, plasmid-mediated, multiresistance (chloramphenicol, erythromycin, streptomycin, and tetracycline) in a clinical isolate of *L monocytogenes* is worrying, especially because the strong evidence that the plasmid originated in "enterococci-streptococci" points to the likelihood of dissemination of resistance to other *L monocytogenes* strains.[17] Since *L monocytogenes* can multiply in mammalian cells, antibiotics that penetrate, concentrate, and retain activity in macrophages are now being considered. Trimethoprim-sulphamethoxazole (co-trimoxazole) given with either rifampicin or ciprofloxacin are potential alternatives to currently used antibiotics.[16]

The incidence of human listeriosis world wide is not known. That it is higher than current data suggest,[4] is almost certain because the clinical features, especially in mild cases, are neither specific nor striking and sporadic cases frequently go undetected; laboratory investigation of the disease varies between and within different countries; statutory notification (usually only of the more overt forms) is required in only a few countries; and voluntary notification is patchy. Current estimates of the incidence of the disease per million of the population vary widely—from 1 to 12.[4,18–23] Although data are difficult to interpret for reasons noted above, there has been a substantial increase in cases of listeriosis in industrialised countries in the past twenty years.

Diagnosis of the illness requires isolation of the organism. Serology is not useful because *L monocytogenes* has several antigens that cross-react with other gram-positive organisms and false-positives may occur.[4,9] Detection of anti-listeriolysin O (a haemolysin) in human sera may prove useful both for serodiagnosis of the disease and for epidemiological studies.[15]

Sources and transmission

L monocytogenes is widespread in the environment. It has been recovered from dust, soil, fresh and salt water, sewage, decaying vegetation, animal feeds (including silage), and fresh and processed foods (including poultry and red meat and meat products, fish, and various fruit and vegetables).[4,5] At least 42 species of wild and domestic animals and 17 avian species, including domestic and game fowl, can harbour the organism and it has been detected in the faeces of both apparently healthy and diseased animals; apparently healthy human faecal carriers are also not uncommon. *L monocytogenes* has been isolated from crustaceans, flies, and ticks.[4] The sources and modes of transmission of most human infections remain unknown, largely because of this ubiquity in the environment and because of the lack, until very recently, of sufficiently discriminating methods for typing of isolates; however, foods are now regarded as an important source.

The rare infections in veterinary practitioners and farmers are associated with the handling of infected material.[4,5] Transplacental transmission from the mother, direct infection in the vaginal canal, and occasional nosocomial infection cause neonatal listeriosis.[4] Possible sources of infection in mothers or non-pregnant adults were, until the early 1980s, not identified. Without proven evidence, and because listeriosis was well known in sheep and cattle, the human disease was regarded as a zoonosis. Contaminated feed, especially poor quality silage, has long been implicated in animal listeriosis.[5] In the large outbreak of human listeriosis in Germany (1949–51), contaminated milk was suspected as the vehicle of infection,[5] but in sub-sequent large and sporadic outbreaks in various countries, contaminated food does not seem to have been considered until recently. Since 1980, there have been four large outbreaks and some sporadic outbreaks of human listeriosis in Europe and North America, in which foods have been clearly implicated as the means of transmission.[4,6] These foods include coleslaw, prepared from cabbages fertilised with sheep manure; milk, almost certainly contaminated after pasteurisation; "Mexican-style" soft cheese, in which some of the milk used to make the cheese had probably not been pasteurised; Vacherin mont d'or cheese ripened in heavily contaminated ripening cellars; other cheeses, vegetable rennet, and cook-chilled chicken; and salted mushrooms.[4,6,24] Large-scale surveys of various products have detected *L monocytogenes* in many types of food, of which uncooked hot dogs, undercooked chicken,[4] and pâté[25] are highly suspect as sources of infection. There is circumstantial evidence that contamination of food materials and production areas probably derives from environmental sources; such evidence also suggests that initial contamination is reinforced by modern methods of feed for animals (eg, greater use of big-bale produced silage[26]) and food production (eg, the development of automated production methods) and increased use of both commercial and domestic refrigeration. Current epidemiological studies may lead us to the main sources of infection and subsequent control of the disease. However, complete eradication of such a ubiquitous organism with the properties to survive and multiply in apparently inhospitable environments is impossible.

Pathogenesis

Knowledge of the pathogenesis of listeriosis remains imperfect. Even though *L monocytogenes* is a facultative, intracellular pathogen, widely distributed in the environment, overt listeriosis is rare. Whereas newborn babies or immunocompromised adults are the most susceptible individuals, infection of apparently healthy people also occurs.[4] Moreover, the organism has been recovered from the faeces of symptom-free individuals, including those at high risk.[4,10] Clearly, exposure to the organism does not always result in disease. The likelihood of systemic infection almost certainly depends on host susceptibility, infectious dose, and virulence factors of the organism.

The importance of T-cell mediated immunity in *L monocytogenes* infection is well documented and accords with the association of listeriosis with immunosuppressive therapy and pregnancy.[4] The role of humoral defences is less well understood. That opsonisation may play a part in the immune response to infection is evidenced by the enhanced susceptibility of newborn babies; this is attributed to the low IgM concentrations and the reduced activity of the classic complement pathway during the neonatal period.[4]

The infectious dose and the main portal of entry for the pathogen are unknown. Immunosuppressed or pregnant mice succumb to smaller infectious doses than do adult mice.[4] Evidence from animal models of listeriosis after ingestion of the organism and from outbreaks of human disease associated with ingestion of contaminated food points to the intestine as an important route. There has been speculation that gastric acid neutralisation by the use of antacids or cimetidine predisposes to infection;[4] it is also possible that in foodborne listeriosis the type of food may be important in avoiding the bactericidal effect of the low pH of the stomach. Damage to the intestinal mucosa due to a pre-existing lesion or a co-infecting microbe has also been suggested, but is unproven, as a predisposing factor.[4] Intestinal ulceration was noted as early as 1927 as one of the main signs of listeriosis induced by feeding contaminated food to gerbils, and a later animal model of listerial enteritis indicated invasion of the epithelial barrier.[4] More recently, in-vitro studies have shown that *L monocytogenes* invades enterocyte-like cells by inducing phagocytosis and that a pre-existing lesion in the epithelial barrier is not required for invasion.[27]

Various products have been considered as virulence determinants in *L monocytogenes*—eg, haemolysin (listeriolysin O); catalase; superoxide dismutase; and the surface components referred to as monocytosis producing activity (MPA), immunosuppressive activity (ISA), the delayed type hypersensitivity protein (DTH), and protein p60. Of these, listeriolysin O is now recognised as an essential virulence factor.[28] Current evidence suggests that protein p60 promotes adhesion and penetration into mammalian cells by inducing phagocytosis.[29] Inside the phagocytes listeriolysin O lyses the membrane-bound phagocytic vacuole (a process essential for survival and multiplication). Once released into the cell cytoplasm, *L monocytogenes* reacts with the host-cell microfilaments and becomes enveloped in a thick coat of F-actin, which facilitates penetration into adjacent cells where listeriolysin O lysis of the membrane-bound phagocytic vacuole is repeated.[30] Non-haemolytic mutants of *L monocytogenes* demonstrate virtually no intracellular spread. The role of the other putative virulence factors has not yet been elucidated.

There has lately been a renewed interest in early reports of the increased virulence of *L monocytogenes* grown at refrigeration temperatures. Virulence of *L monocytogenes* is increased at 4°C for intravenously but not orally infected mice.[31] This increased virulence is almost certainly due to production of stress proteins and reflects a highly artificial situation in that it requires direct entry of the organism into the blood stream. The same studies do not support claims that haemolysin production is enhanced at 4°C. Thus, there is no evidence to support the notion that *L monocytogenes* in refrigerated products has enhanced virulence when ingested.

Epidemiological data have been interpreted as evidence that *L monocytogenes* serovars 1/2a, 1/2b, and especially 4b are more virulent than other serovars.[32] However, an association between virulence and antigenic composition in *L monocytogenes* is unlikely; moreover, serovars 1/2a, 1/2b, and 4b occur also in *L seeligeri*. As judged by MEE typing of clinical and food isolates, there are two distinct groups of electrophoretic types (ETs) of *L monocytogenes*—strains with flagellar antigen a and strains with flagellar antigen b.[13] Although there are no obvious trends in ET distribution between geographic areas, type of clinical presentation, or severity of disease within the two groups, one ET (containing serovar 4b isolates) may be an especially virulent clone.[13]

Further work is needed on the pathogenesis and transmission of *L monocytogenes*. In the meantime, control of listeriosis is best achieved by an awareness of the ubiquity of the organism and especially of those environments that favour its multiplication. Thus, the collaboration of medical, veterinary, and food bacteriologists; farmers; food manufacturers; technologists; and engineers is essential: the decrease in the number of reported cases in the past year in industrialised countries shows that this is already happening.

References

1. Murray EGD, Webb RA, Swann MBR. A disease of rabbits characterised by a large mononuclear leucocytosis, caused by a hitherto undescribed bacillus *Bacterium monocytogenes* (n. sp.) *J Pathol Bacteriol* 1926; **29**: 407–39.
2. Rocourt J, Wehmeyer U, Cossart P, Stackebrandt E. Proposal to retain *Listeria murrayi* and *Listeria grayi* in the genus *Listeria*. *Int J Syst Bacteriol* 1987; **37**: 298–300.
3. Seeliger HPR, Jones D. Genus *Listeria*. In: Sneath PHA, Mair NS, Sharpe ME, Holt JG, eds. Bergey's manual of systematic bacteriology. Vol 2. Baltimore: Williams and Wilkins, 1986: 1235–45.
4. Gellin BG, Broome CV. Listeriosis. *JAMA* 1989; **261**: 1313–20.
5. Gray ML, Killinger AH. *Listeria monocytogenes* and listeric infections. *Bacteriol Rev* 1966; **30**: 309–82.
6. Lund BM. The prevention of foodborne listeriosis. *Br Food J* 1990; **92**: 13–22.
7. Farber JM, Sanders GW, Dunfield S, Prescott R. The effect of various acidulants in the growth of *Listeria monocytogenes*. *Letts Appl Microbiol* 1989; **9**: 181–83.
8. Lamont RJ, Postlethwaite R, MacGowan AP. *Listeria monocytogenes* and its role in human infection. *J Infect* 1988; **17**: 7–28.
9. Seeliger HPR, Höhne K. Serotyping of *Listeria monocytogenes* and related species. In: Bergan T, Norris JR, eds. Methods in microbiology. Vol 13. London: Academic Press, 1979: 31–49.
10. McLauchlin J. A review: *Listeria monocytogenes*, recent advances in the taxonomy and epidemiology of listeriosis in humans. *J Appl Bacteriol* 1987; **63**: 1–11.
11. McLauchlin J, Audurier A, Taylor AG. The evaluation of a phage typing system for *Listeria monocytogenes* for use in epidemiological studies. *J Med Microbiol* 1986; **22**: 357–65.
12. Piffaretti JC, Kressebuch H, Aeschbacher M, et al. Genetic characterization of clones of the bacterium *Listeria monocytogenes* causing epidemic disease. *Proc Natl Acad Sci USA* 1989; **86**: 3818–22.
13. Bibb WF, Gellin BG, Weaver R, et al. Analysis of clinical and food-borne isolates of *Listeria monocytogenes* in the United States by multilocus enzyme electrophoresis and application of the method to epidemiological investigations. *Appl Environ Microbiol* 1990; **56**: 2133–41.
14. Saunders NA, Ridley AM, Taylor AG. Typing of *Listeria monocytogenes* for epidemiological studies using DNA probes. *Acta Microbiol Hung* 1989; **36**: 205–09.
15. Berche P, Reich KR, Bonnichon M, et al. Detection of anti-listeriolysin O for serodiagnosis of human listeriosis. *Lancet* 1990; **335**: 624–27.
16. Boisivon A, Guiomar C, Carbon C. *In vitro* bactericidal activity of amoxicillin, gentamicin, rifampicin, ciprofloxacin and trimethoprim- sulfamethoxazole alone or in combination against *Listeria monocytogenes*. *Eur J Clin Microbiol Infect Dis* 1990; **9**: 206–09.
17. Poyart-Salmeron C, Carlier C, Trieu-Cuot P, Courtieu A-L, Courvalin P. Transferable plasmid-mediated antibiotic resistance in *Listeria monocytogenes*. *Lancet* 1990; **335**: 1422–26.
18. Broome CV, Gellin B, Schwartz B. Epidemiology of listeriosis in the United States. In: Miller AJ, Smith JL, Somkuti GA, eds. Foodborne listeriosis. Amsterdam: Elsevier, 1990: 61–65.
19. Campbell DM. Human listeriosis in Scotland 1967–1988. *J Infect* 1990; **20**: 241–50.
20. McLauchlin J. Human listeriosis in Britain, 1967–85: a summary of 722 cases. 1. Listeriosis during pregnancy and in the newborn. *Epidemiol Infect* 1990; **104**: 181–89.

21. McLauchlin J. Human listeriosis in Britain, 1967–85: a summary of 722 cases. 2. Listeriosis in non-pregnant individuals, a changing pattern of infection and seasonal influence. *Epidemiol Infect* 1990; **104**: 191–201.
22. Samuelsson S, Rothgardt NR, Carvajal A, Frederiksen W. Human listeriosis in Denmark 1981–87 including an outbreak Nov 1985–March 1987. *J Infect* 1990; **20**: 251–59.
23. Varughese PV, Carter MD. Human listeriosis surveillance in Canada—1988. *Can Dis Weekly Rec* 1989; **15**: 213–20.
24. Junttila J, Brander M. *Listeria monocytogenes* septicaemia associated with consumption of salted mushrooms. *Scand J Infect Dis* 1989; **21**: 339–42.
25. Morris IJ, Ribeiro CD. *Listeria monocytogenes* and pâté. *Lancet* 1989; ii: 1285–86.
26. Fenlon DR. Wild birds and silage as reservoirs of *Listeria* in the agricultural environment. *J Appl Bacteriol* 1985; **59**: 537–43.
27. Gaillard JL, Berche P, Mounier J, Richard S, Sansonnetti PJ. In vitro model of penetration and intracellular growth of *Listeria monocytogenes* in the human enterocyte-like cell line Caco-2. *Infect Immun* 1987; **55**: 2822–29.
28. Cossart P, Vicente MF, Mengard J, Baquero F, Perez-Diaz JC, Berche P. Listeriolysin O is essential for virulence of *Listeria monocytogenes*:direct evidence obtained by gene complementation. *Infect Immun* 1989; **57**: 3629–36.
29. Köhler S, Leimeister-Wachter M, Chakraborty T, Lottspeich F, Goebel W. The gene coding for protein p60 of *Listeria monocytogenes* and its use as a specific probe for *Listeria monocytogenes*. *Infect Immun* 1990; **58**: 1943–50.
30. Mounier J, Ryter A, Coquis-Rondon M, Sansonetti PJ. Intracellular and cell to cell spread of *Listeria monocytogenes* involves interaction with F-actin in the enterocyte cell line Caco-2. *Infect Immun* 1990; **58**: 1048–58.
31. Czuprynski CJ, Brown JF, Roll JT. Growth at reduced temperatures increases the virulence of *Listeria monocytogenes* for intravenously but not intragastrically inoculated mice. *Microb Pathogen* 1989; **7**: 213–23.
32. McLauchlin J. Distribution of serovars of *Listeria monocytogenes* isolated from different categories of patient with listeriosis. *Eur J Clin Microbiol Infect Dis* 1990; **9**: 210–14.

11
Pathogenic *Escherichia coli*, *Yersinia enterocolitica*, and *Vibrio parahaemolyticus*

Michael P. Doyle

Pathogenic *Escherichia coli*

Before the recognition of *E coli* O157:H7 as a cause of foodborne disease in 1982, pathogenic *E coli* were regarded as infrequent causes of food-associated illness in developed countries. However, since then, *E coli* O157:H7 has been responsible for many beef-associated and raw milk-associated cases of haemorrhagic colitis or haemolytic uraemic syndrome in the USA and Canada. Food-associated outbreaks of gastroenteritis due to other types of pathogenic *E coli* in developed countries are rare compared with most other recognised foodborne enteric pathogens.

By contrast, in developing countries pathogenic *E coli*, with the possible exception of *E coli* O157:H7, are important causes of gastrointestinal disease, especially among infants and children. Contaminated water is a principal vehicle for transmission of these infections, either by direct consumption or by its presence in foods that are irrigated, washed, or prepared with such water. Furthermore, travellers' diarrhoea, which is the commonest health disorder of travellers to developing countries, is largely due to enterotoxigenic *E coli* acquired by the ingestion of faecally contaminated food or water.

E coli is part of the normal microflora of the intestinal tract of human beings and of most warm-blooded animals. Generally, the strains that colonise the human bowel are harmless commensals. However, there are pathogenic strains that cause distinct syndromes of diarrhoeal disease and that have been associated with foodborne illness. These foodborne pathogens are grouped into the following four categories based on distinct virulence properties, different interactions with the intestinal mucosa, distinct clinical syndromes, differences in epidemiology, and distinct O:H serogroups.[1]

ENTEROPATHOGENIC *E COLI* (EPEC)

EPEC include *E coli* of specific serotypes that historically have been associated with outbreaks of infantile diarrhoea, but whose pathogenic mechanisms are not related to heat-labile or heat-stable enterotoxins, to shigella-like invasiveness, or to verotoxin.[2-4] Serogroups of EPEC include O18ab, O18ac, O26, O44, O55, O86, O111, O114, O119, O125, O126, O127, O128ab, O142, and O158. The organism is best known as a cause of outbreaks of diarrhoea, which often occur in hospital nurseries. Outbreaks of EPEC diarrhoea are now uncommon in countries with good hygienic standards; however, EPEC are an important cause of infantile diarrhoea in many developing countries.

A few foodborne and waterborne outbreaks of EPEC gastroenteritis have been reported but not recently.[2] In these outbreaks likely sources of contamination were foodhandlers or water supplies that had been contaminated by sewage. Acquisition of immunity may explain the rare occurrence of EPEC illness and the high frequency of EPEC carriers among adults.

The mechanism of pathogenicity of EPEC is not yet fully defined, but adherence to the intestinal mucosa is believed to be an important factor for colonisation of the intestinal tract. In-vivo studies have shown that EPEC strains possess a specific type of attaching and effacing adherence to intestinal mucosal epithelial cells.[4,5] The bacteria destroy microvilli without further invasion and adhere closely to the intestinal cell membrane, with the membrane partly enveloping each bacterium. This same type of lesion has been observed in biopsy samples of intestinal tissue from infants with EPEC infections and is postulated to be responsible for the diarrhoea associated with such infections.

ENTEROINVASIVE *E COLI* (EIEC)

EIEC produce an invasive dysentery type of diarrhoeal illness in man. Serogroups principally associated with EIEC infection are O28ac, O29, O124, O136, O143, O144, O152, O164, and O167. Since EIEC were first recognised as a cause of diarrhoea in the mid-1940s, several food-associated outbreaks of EIEC diarrhoea have been reported.[2] There was a major outbreak involving at least 387 individuals in the USA in 1971, which was associated with consumption of imported Brie or Camembert cheese. Investigation of the factory that produced the cheese revealed that equipment for filtering river water used to clean the factory was malfunctioning when the contaminated cheese was being produced. Although EIEC is an established foodborne pathogen, it is an infrequently reported cause of illness in most developed countries.

For most food-related outbreaks, sources of EIEC contamination are associated with an infected foodhandler or contact with water that has probably been contaminated with sewage. Several lines of evidence point to infected human beings as the principal reservoir of EIEC—eg, outbreaks of

EIEC diarrhoea attributed to person-to-person transmission (presumably the faecal-oral route), the organism's lack of association with animals and with foods of animal origin, and the similar properties of EIEC to *Shigella*, which is carried principally by infected human beings.[2]

The pathogenesis of EIEC diarrhoea resembles that of *Shigella* dysentery: epithelial invasion and intracellular multiplication in the colon lead to inflammation and ulceration of the mucosa.[6] The colon is the predominant site of bacterial invasion. The invasive ability of EIEC is associated with a plasmid that encodes for several outer membrane polypeptides involved in invasiveness. These polypeptides are antigenically closely related, if not identical, to those of *Shigella*.

ENTEROTOXIGENIC *E COLI* (ETEC)

ETEC are the most common cause of travellers' diarrhoea for individuals who travel from areas of good hygiene and temperate climate to areas with lower hygienic standards, particularly developing countries in the tropics.[2] Although ETEC uncommonly cause illness in developed countries with good hygiene, they are an important cause of diarrhoea in all age groups (especially in infants and young children) in developing countries and tropical areas of poor hygiene. A child who lives in an area of poor hygiene may have two or three ETEC infections a year during the first 2–3 years of life. The most common serogroups of ETEC include O6, O8, O15, O20, O25, O27, O63, O78, O80, O85, O115, O128ac, O139, O148, O153, O159, and O167.

ETEC infection is acquired principally by ingestion of contaminated food or water. Foodborne or waterborne outbreaks of ETEC infection in developed countries are rare. A waterborne outbreak which involved more than 2000 individuals at a US national park in 1975 was caused by drinking water that was contaminated with raw sewage. Infected foodhandlers are often implicated as the source of contamination of foods associated with ETEC outbreaks and it is presumed that human beings are the major reservoir of ETEC. In developing countries, ETEC are often present in the faeces of symptom-free human carriers.[2]

The mechanism of pathogenicity of ETEC is in many respects similar to that of *Vibrio cholerae* group O1.[2] After ingestion, ETEC cells that have survived the hostile environment of the stomach penetrate the mucous layer of the proximal small intestine where they adhere to mucosal cells and elaborate heat-labile (LT) or heat-stable (ST) enterotoxins or both. LT is immunologically related to cholera enterotoxin; ST is non-antigenic and has a low molecular weight. LT stimulates adenylate cyclase and ST stimulates guanylate cyclase of the enterocyte leading to the accumulation of cyclic AMP and cyclic GMP, respectively. These cyclic nucleotides cause fluid secretion, which results in watery diarrhoea. The ability to produce enterotoxin alone is not sufficient for ETEC to produce diarrhoea; the bacteria must also colonise the mucosal surface of the small intestine, which is accomplished by adhesive factors composed of non-flagellar filamentous fimbriae. These fimbriae are host-specific, and bind to specific receptors in the membrane of cells of the small intestine. Hence, fimbriae that colonise man do not seem to colonise animals and vice versa. ETEC can produce LT and possibly ST in milk, but only under extreme conditions of temperature abuse. Edible foods are unlikely to contain pre-formed ETEC enterotoxin, so concerns about foods containing such toxins are unfounded.

Outbreaks of ETEC and EIEC diarrhoea caused by contaminated Camembert cheese have prompted studies to assess the survival and growth characteristics of these organisms in this cheese:[2] pathogenic *E coli* generally do not survive well and die off during storage in cheese made with contaminated milk. However, when the organisms contaminate the mold that grows on the cheese surface during ripening, *E coli* grow rapidly and large numbers survive during extended storage. Deamination of proteins by the proliferating mold increases the surface pH of the cheese to near neutrality, which makes the conditions favourable for *E coli*. Hence, growth and long-term survival of pathogenic *E coli* on Camembert cheese largely depend on the stage during cheese production when the product is contaminated. Like most *E coli*, pathogenic *E coli* do not survive well in fermented dairy products at pH 5 or below.

ENTEROHAEMORRHAGIC *E COLI* (EHEC)

EHEC include strains of *E coli* that have been associated with haemorrhagic colitis and that produce one or more verotoxins. So far, only one serotype of EHEC—namely *E coli* O157:H7—has been associated with foodborne disease.[2] This organism is now recognised as an important cause of foodborne disease, with outbreaks being reported in the USA, Canada, and the UK. The severe features of illness, including haemorrhagic colitis, haemolytic uraemic syndrome, and thrombotic thrombocytopenic purpura, characterise it as a more important pathogen than the other types of "enteropathogenic" *E coli*.

Most outbreaks of *E coli* O157:H7 infection have been linked to consumption of undercooked ground beef, and, to a lesser extent, to the drinking of unpasteurised milk. In surveys of retail raw meats and poultry, 3·5% of ground beef, 1·5% of pork, 1·5% of poultry, and 2·0% of lamb were contaminated with *E coli* O157:H7.[7] The organism was also isolated from venison implicated in a case of haemorrhagic colitis. Dairy cattle, especially young animals within herds, is a principal reservoir of the organism. The organism has been isolated from the faeces of dairy calves and heifers, including calves 1–3 weeks old with *E coli* bacillosis. Since *E coli* O157:H7 has been isolated from meats of several animal species, it is possible that animals other than cattle may be reservoirs of the organism. Although many outbreaks of *E coli* O157:H7 infection have been linked to consumption of cooked ground beef, the organism is not unusually heat resistant.[2] Heating ground beef sufficiently to kill typical strains of salmonellae will also kill *E coli* O157:H7. Thus, in outbreaks associated with consumption of this food, the meat was probably undercooked.

The mechanism of pathogenicity for *E coli* O157:H7 has not yet been fully elucidated, but important virulence factors have been identified. All clinical isolates produce one or two verotoxins.[3] These toxins are cytotoxic to Vero cells (African green monkey kidney) grown in tissue culture. One of the toxins (VT-1) is immunologically and structurally indistinguishable from Shiga toxin which is produced by *Sh dysenteriae* type 1. Hence, these verotoxins are also known as Shiga-like toxins. The toxins are cytotoxic to the colon of mice and rabbits. It is hypothesised that *E coli* O157:H7 colonises the intestinal tract and elaborates its toxin(s), which subsequently act on the colon. The organism can adhere to human small intestine cells in tissue culture.

IMPORTANCE OF PATHOGENIC *E COLI* AS FOODBORNE PATHOGENS
(Table 11.1)

Symptomatic and symptom-free human carriers are presumed to be the principal reservoir of EPEC, EIEC, and ETEC strains that cause human illness. These bacteria are present in the intestinal tract of carriers and are excreted in their faeces. Infected foodhandlers with poor personal hygiene or water contaminated by human sewage are sources of food contamination. Control measures to prevent food-poisoning include education of food workers about safe foodhandling techniques and proper personal hygiene, properly heating foods to kill *E coli*, and holding foods under appropriate conditions to avoid bacterial multiplication. Additionally, untreated human sewage should not be used to irrigate or fertilise vegetables and crops; likewise, untreated surface water should not be used to clean food processing equipment and food contact surfaces.

The principal reservoir of *E coli* O157:H7 is the intestinal tract of dairy cattle and perhaps other animals used in the production of food. Therefore, raw foods of animal origin (especially beef and cow's milk) may be contaminated with the organism via faecal contact during slaughter or milking procedures. The use of good manufacturing practices in the processing of foods of animal origin and proper heating of foods before consumption are important control measures for the prevention of *E coli* O157:H7 infections.

Yersinia enterocolitica

Although identified as a human pathogen in 1939, *Y enterocolitica* was not

Table 11.1   Characteristics of pathogenic *E coli* associated with foodborne disease

Type of *E coli*	Known or presumed reservoir	Sources of food contamination	Food vehicles of outbreaks
EPEC	Man	Foodhandlers, sewage	Coffee substitute, cold port, meat pie, water
EIEC	Man	Foodhandlers, sewage	Brie and Camembert cheese, canned salmon, potato salad, water
ETEC	Man	Foodhandlers, sewage	Brie cheese, curried turkey mayonnaise, prepared food (restaurant cafeteria, and cruise ship), water
EHEC	Dairy cattle	Cattle faeces, meat handling facilities, dairies	Undercooked ground beef, unpasteurised milk

recognised as a foodborne pathogen until the mid-1970s. Since then, a few large food-associated outbreaks of yersiniosis have been reported, mainly in the USA; however, sporadic yersiniosis is uncommon in that country.[8]

The true incidence of yersiniosis is not known, but infection is most prevalent in the cooler regions of Europe and North America. In some countries, such as Belgium, Canada, the Netherlands, Australia, and parts of Germany, *Y enterocolitica* has surpassed *Shigella* and rivals *Salmonella* as a cause of acute gastroenteritis.[8,9] Interestingly, there is seasonal variation in the occurrence of yersiniosis; most cases are reported during the autumn and winter months.[8,10] Additionally, different serotypes of *Y enterocolitica* are associated with human infections in different regions of the world. In Europe, Canada, and Japan, sporadic infections caused by serovar O:3, and to a lesser extent O:9, are predominant. In the USA, sporadic cases are associated with multiple serovars and are uncommon, but most reported outbreaks have been caused by serovar O:8.[8,9,11]

Y enterocolitica is often found in the environment[11] and in the alimentary tract of various animals[8,11] but, with the exception of strains from pigs,[8–11] most isolates are types not associated with human infection and are usually non-pathogenic. Pigs are regarded as major reservoirs of pathogenic *Y enterocolitica* since the serovars most commonly involved in human infections (ie, O:3, O:5,27, O:8, O:9) are often carried in the oral cavity or intestinal tract of healthy pigs.[9,11] Many investigators believe that pathogenic types of *Y enterocolitica* are normal residents of the oral cavity of pigs and that these animals have a major role in the epidemiology of human infections.[8–11]

Contaminated pork, foods that contact raw pork, or swine wastes are likely vehicles for sporadic cases of yersiniosis. In a case-control study to determine the risk factors for *Y enterocolitica* infection in Belgium, which has a high incidence of yersiniosis, the disease was strongly associated with eating raw pork in the two weeks before illness.[10] A large outbreak due to an unusual serovar (O:13a,13b) of *Y enterocolitica* was associated with consumption of pasteurised milk.[11] The probable source of the organism was pigs whose manure contaminated crates used to transport both outdated milk to a pig farm and fresh milk to retail outlets. Contaminated spring water and stream water also have been identified as vehicles of yersiniosis.[11] A large outbreak in the USA was attributed to consumption of tofu packed in untreated spring water that was contaminated with *Y enterocolitica* serovar O:8.[11]

Y enterocolitica is widely occurring and is often present in foods of animal origin. The organism has been isolated from beef, lamb, poultry, pork, milk, crabs, oysters, and shrimp. However, with the exception of pork, isolates obtained from most foods are non-pathogenic. Pathogenic strains have been isolated from pork cuts and, with high frequency, from porcine tongues.[10,11]

Unlike most enteric pathogens, *Y enterocolitica* is a psychrotroph that can grow at 0–2°C. Although growth is moderately slow at refrigeration temperatures, the organism can grow to large numbers in raw or cooked meat during extended refrigerated storage; in one study,[12] a few hundred *Y enterocolitica* in raw pork held at 7°C grew to more than 10 cells/g within ten days. The organism is destroyed by standard pasteurisation treatments and by thorough cooking.

Virulence of *Y enterocolitica* is associated with a 40–48 MD plasmid that encodes for several virulence-related antigens.[8,11] Strains cured of this plasmid are no longer virulent. Several temperature-dependent characteristics of virulent strains, such as calcium-dependent growth, serum resistance, expression of unique proteins and antigens, and autoagglutination in broth culture, are associated with the virulence plasmid.[8,11] There are two types of virulent *Y enterocolitica*.[11] When given by mouth to mice, one type (eg, serovar O:8 and O:21 strains) produces fatal infections, whereas the other (eg, serovar O:3, O:5,27, and O:9 strains) colonises the mouse intestines which leads to long-term faecal excretion and diarrhoeal symptoms but without fatal consequences. The course of infection involves invasion of the mucosa of the ileum, followed by multiplication of yersiniae within Peyer's patches.[8,11] The bacteria then migrate to the mesenteric lymph nodes, from which systemic infection can arise. As soon as yersiniae have passed through the intestinal epithelium, they are enveloped by phagocytic cells, such as macrophages.[13] The organisms reside within vacuoles of these cells and seem to multiply within them.

Most clinical isolates of *Y enterocolitica* produce a heat-stable enterotoxin that in many ways is similar to the heat-stable toxin of *E coli*. However, because most *Y enterocolitica* do not produce enterotoxin in vitro at temperatures greater than 30°C and because strains that do not produce enterotoxin in vitro have caused diarrhoea in experimentally infected mice, this enterotoxin probably does not play an important part in *Y enterocolitica* infection.[8,11]

IMPORTANCE AS A FOODBORNE PATHOGEN

Since *Y enterocolitica* is a psychrotroph, cold storage is ineffective for control of its growth in foods. Consequently, measures must be taken to prevent contamination of foods or to treat foods with heat or some other means of microbial inactivation to destroy the organism. That pathogenic strains are seldom present in most foods may explain why food-associated outbreaks of yersiniosis are not more common.

Vibrio parahaemolyticus

Japanese investigators in 1950 were the first to identify *V parahaemolyticus* as a cause of foodborne disease. The organism is now recognised as the principal cause of foodborne outbreaks in Japan, and as a health hazard in seafoods consumed world wide.[14]

V parahaemolyticus is a halophilic microorganism—ie, it requires sodium chloride for growth. Because of its halophilic nature, the organism is widely distributed in marine environments but is seldom associated with freshwater or non-marine environments. The organism is frequently present in fish and shellfish caught in estuaries and inshore coastal waters.[14,15] There is a pronounced seasonality in the prevalence of *V parahaemolyticus* in the estuarine environment, with the greatest frequency of isolation and highest vibrio concentrations during the summer to early autumn months.[14,15] Isolation of the organism from shellfish, sediment, and water is infrequent or rare when water temperatures fall below 13–15°C. The organism's affinity for chitin, which is a constituent of the exoskeleton of many plankton species, is an important factor that influences the survival of *V parahaemolyticus* during winter.[14] The organism adsorbs to chitinous material of plankton, which sinks to the estuarine bottom during the winter. As the temperature increases the following year, the organism reappears in the water column as part of the annual cycle of sedimentation and resuspension.

Gastroenteritis due to *V parahaemolyticus* is nearly always associated with consumption of contaminated seafood and shellfish.[15] Volunteer studies have shown that the minimum infectious dose of virulent strains of *V parahaemolyticus* ranges from 10^5 to 10^7 cells. Usually the numbers of the organism on freshly caught seafoods are low—ie, up to 10^2/g. However, in market shellfish, especially during summer, counts of 10^3/g are not uncommon; in Japan counts ranging from 10^3/g to 10^4/g have been reported.[14] Proper refrigeration of seafoods, both before and after cooking, is essential to control the growth of vibrios. The organism has a remarkably short generation time in seafood that has been temperature abused—eg, in raw squid and horse mackerel, 13 min at 37°C and 15–18 min at 30°C, respectively, and in boiled octopus, 12 min at 30°C.

Seafoods should be properly cooked to kill vibrios. Food-poisoning has resulted from consumption of raw seafood dishes prepared in traditional Japanese culinary style and from insufficiently cooked seafoods. Heating seafood at 60°C for 15 min will kill *V parahaemolyticus*.[14] Additionally, the organism is sensitive to cold storage—eg, a 1-log_{10} to 4-log_{10} decrease of vibrios in raw fish held at 4°C or frozen (-2°C, -10°C, or -16°C).[14] Hence, refrigerated storage of seafoods does more than merely prevent the growth of *V parahaemolyticus*.

The mechanism of pathogenicity is unknown. However, there is a strong correlation between pathogenicity of *V parahaemolyticus* and possession of a thermostable direct haemolysin that produces beta haemolysis of human erythrocytes in Wagatsuma agar.[14,15] This reaction is known as the Kanagawa phenomenon. Almost all strains isolated from patients with gastroenteritis are Kanagawa-positive. These strains produce fluid accumulation in rabbit ligated ileal loops, whereas Kanagawa-negative strains do not. Interestingly, most isolates of *V parahaemolyticus* from foods and marine environments are Kanagawa-negative.

IMPORTANCE AS A FOODBORNE PATHOGEN

The widespread presence of *V parahaemolyticus* in estuarine waters suggests that seafood harvested from such water, especially in the summer and early autumn, will be contaminated with the organism.[14] Fortunately, most strains in seafoods are Kanagawa-negative and are likely to be avirulent. However, if present, virulent strains can grow to large numbers during even short periods of inadequate refrigerated storage. Foodborne outbreaks of *V parahaemolyticus* gastroenteritis occur infrequently in the USA and Europe but are common in Japan. These outbreaks have usually happened because seafoods were under-refrigerated, inadequately cooked, or recontaminated and temperature abused after cooking. Proper cooking and refrigeration of seafoods are important factors in the prevention of *V parahaemolyticus* gastroenteritis.

References

1. Levine MM. *Escherichia coli* that cause diarrhea: enterotoxigenic, enteropathogenic, enteroinvasive, enterohemorrhagic, and enteroadherent. *J Infect Dis* 1987; **155**: 377–89.
2. Doyle MP, Padhye VV. *Escherichia coli*. In: Doyle MP, ed. Foodborne bacterial pathogens. New York: Marcel Dekker, 1989: 236–81.
3. Karmali MA. Infection by verocytotoxin-producing *Escherichia coli*. *Clin Microbiol Rev* 1989; **2**: 15–38.
4. Robins-Browne RM. Traditional enteropathogenic *Escherichia coli* of infantile diarrhea. *Rev Infect Dis* 1987; **9**: 28–53.
5. Moon HW, Whipp SC, Arzenio RA, Levine MM, Gianella RA. Attaching and effacing activities of rabbit and human enteropathogenic *Escherichia coli* in pig and rabbit intestines. *Infect Immun* 1983; **41**: 1340–51.
6. Gross RJ, Rowe B. *Escherichia coli* diarrhoea. *J Hyg* 1985; **95**: 531–50.
7. Doyle MP, Schoeni JL. Isolation of *Escherichia coli* O157:H7 from retail fresh meats and poultry. *Appl Microbiol* 1987; **53**: 2394–96.
8. Cover TL, Aber RC. *Yersinia enterocolitica*. *N Engl J Med* 1989; **321**: 16–24.
9. Doyle MP. Food-borne pathogens of recent concern. *Ann Rev Nutr* 1985; **5**: 25–41.
10. Tauxe RV, Vandepitte J, Wauters G, et al. *Yersinia enterocolitica* infections and pork: the missing link. *Lancet* 1987; i: 1129–32.
11. Scheimann DA. *Yersinia enterocolitica* and *Yersinia pseudotuberculosis*. In: Doyle MP, ed. Foodborne bacterial pathogens. New York: Marcel Dekker, 1989: 601–72.
12. Hanna MO, Stewart JC, Zink DL, Carpenter ZL, Vanderzant. Development of *Yersinia enterocolitica* on raw and cooked beef and pork at different temperatures. *J Food Protect* 1977; **42**: 1180–84.
13. Finlay BB, Falkow S. Common themes in microbial pathogenicity. *Microbiol Rev* 1989; **53**: 210–30.
14. Twedt RM. *Vibrio parahaemolyticus*. In: Doyle MP, ed. Foodborne bacterial pathogens. New York: Marcel Dekker, 1989: 543–68.
15. West PA. The human pathogenic vibrios: a public health update with environmental perspectives. *Epidemiol Infect* 1989; **103**: 1–34.

12
Foodborne disease due to *Bacillus* and *Clostridium* species

Barbara M. Lund

Bacillus species

OUTBREAKS OF FOODBORNE DISEASE DUE TO *B CEREUS*

Outbreaks of food-poisoning due to *Bacillus* spp have been described since the beginning of this century, but it was in the early 1950s, after the taxonomy of *Bacillus* had been clarified, that *B cereus* was recognised as an important food-poisoning bacterium world wide.[1] Because of the limitations of reporting systems of foodborne disease it is difficult to compare the incidence of outbreaks in different countries. In the past 30 years, however, the incidence of outbreaks due to *B cereus* has been especially high in the Netherlands, Finland, Hungary, and Canada. From 1986 to 1988 there has been a more than six-fold increase in the numbers of reported cases of food-poisoning due to *Bacillus* spp in England and Wales (Public Health Laboratory Service [PHLS] Communicable Disease Surveillance Centre, unpublished) (Table 12.1); *B cereus* accounted for most of these cases. The number of cases reported in 1988 was the highest annual total since records of this illness were first kept in 1971.

B cereus causes two distinct forms of gastroenteritis—namely, the "diarrhoeal syndrome" (associated with proteinaceous foods, vegetables, sauces, and puddings), and the "emetic syndrome" (associated with farinaceous foods, particularly cooked rice). The diarrhoeal syndrome was first described in 1950, following four outbreaks in Norway. In outbreaks of the diarrhoeal form of food-poisoning, the levels of *B cereus* found in the implicated foods are usually in the range $5 \times 10^5 - 9 \cdot 5 \times 10^8$ colony forming units (cfu)/g. The emetic form of the disease was first identified in the UK in the early 1970s. Many of the outbreaks in the UK have been associated with cooked (particularly fried) rice from Chinese restaurants and take-away outlets. The occurrence of incidents is linked to the practice of saving portions of boiled

Table 12.1   Food-poisoning due to *Bacillus* spp *Clostridium* spp, England and Wales, 1986–88*

	No of cases			
	1986	1987	1988	Total
Bacterial food-poisoning and *Salmonella* infection†	15 214	18 573	24 941	58 729
Clostridium perfringens	896	1266	1312	3474
Bacillus spp	65	137	418	620
Clostridium botulinum	0	1	0	1
Gastroenteritis due to *Campylobacter*	24 809	27 310	28 761	80 232

Other bacterial food-poisoning and *Campylobacter* infections for comparison.
*From the PHLS Communicable Disease Surveillance Centre, unpublished.
†Excluding gastroenteritis due to *Campylobacter*.

rice from bulk cooking. The boiled rice is allowed to "dry off" at ambient temperature, after which it may be stored overnight or longer before it is fried quickly with beaten egg. These practices result in the survival and proliferation of strains of *B cereus*, originally present in the raw rice, that produce spores with the greatest heat resistance. When the cooked rice is maintained at ambient temperature the spores germinate and rapid growth of the vegetative bacteria occurs. Levels of *B cereus* in foods incriminated in incidents of the emetic form of food-poisoning have ranged from $1 \cdot 0 \times 10^3$ to $5 \cdot 0 \times 10^{10}$ cfu/g; high numbers are frequently also present in faecal samples from patients.

SYMPTOMS OF FOOD-POISONING DUE TO *B CEREUS*

In the diarrhoeal form of infection, the incubation period is from 8 to 16 h (on average 10–12 h); symptoms, which include abdominal pain, profuse watery diarrhoea, rectal tenesmus, and occasionally nausea and vomiting, generally resolve within 12–24 h. In the emetic form, the incubation period is 1–5 h and the symptoms include nausea, vomiting, and malaise (in some cases followed by diarrhoea) which lasts 6–24 h. These two syndromes are believed to be due to the effects of two distinct enterotoxins (Table 12.2).

Most reported cases of *B cereus* food-poisoning in the UK have involved the emetic syndrome,[1] in which the toxin is preformed in the food. In the diarrhoeal form of the disease the enterotoxin is both preformed in food and formed by the bacterium in the intestine. Whether the preformed toxin is sufficiently stable to cause the disease after ingestion has not yet been established (J. Kramer, personal communication).

PROPERTIES OF *B CEREUS*

B cereus is widespread in the environment and is present in most raw foods.[1] Numbers are especially high in some samples of spices and cereals. The high incidence of *B cereus* infection in Hungary from 1960 to 1968, when it ranked as the third most common type of food-poisoning, has

Table 12.2   Characteristics of foodborne disease due to species of *Bacillus* and *Clostridium* and properties of the bacteria*

Characteristic	B cereus		B subtilis	B licheniformis	B pumilus	C perfringens type A	C botulinum	
	Diarrhoeal syndrome	Emetic syndrome					Proteolytic types A, B, F	Non-proteolytic types B, E, F
Onset of symptoms (h)	8–16	1–5	0·17–14	2–14	0·25–11	8–24	2 h–8 days	
Duration of symptoms (h)	12–24	6–24	1·5–8	6–24	NR	12–24	Few days to several months	
Most common symptoms	Abdominal pain, profuse watery diarrhoea, rectal tenesmus	Nausea vomiting, malaise	Vomiting, diarrhoea	Diarrhoea, vomiting, abdominal pain	Diarrhoea, vomiting	Diarrhoea, severe abdominal pain	Nausea, vomiting (may not be due to botulinum toxins), disturbance of vision, dysphagia, generalised weakness, diplopia, dizziness or vertigo	
Less common symptoms	Nausea	Diarrhoea	Abdominal pain, nausea			Nausea	Abdominal pain, cramps, fullness, diarrhoea, urinary retention or incontinence, sore throat, constipation	
Unusual symptoms	Vomiting		Headaches, flushing, sweating					
No of bacteria in incriminated food (/g)	$5 \times 10^5 - 9\cdot5 \times 10^8$	$1 \times 10^3 - 5 \times 10^{10}$	$1 \times 10^5 - 10^9$	$> 10^6$	"High"	$> 10^5$	Usually high (except in infant botulism)	
Toxin formed mainly in food	+†	+	NR	NR	NR	–	+	
Toxin formed mainly in intestine	+†	–	NR	NR	NR	+	Only in infant botulism and rarely in adults	
Requirement of bacteria for oxygen	Aerobe/facultative anaerobe		Aerobe	Aerobe/facultative anaerobe	Aerobe	Anaerobe	Anaerobe	
Temperature range for growth (°C)	5–50		NR	?–55	5–50	15–50	> 10–50	3·3–45
pH range for growth	4·3–9·0		NR	NR	NR	About 5·0–8·3	$> 4\cdot6$–between 8 and 9	

*Compiled mainly from refs 1–3.
†Toxin formed both in food and in intestine (see text).
NR = Not reported.

been attributed to well-spiced meat dishes. The spices often contained high numbers of spores, some of which would survive cooking. Lack of proper refrigeration of meals after cooking would allow surviving spores to germinate and multiply in the food.

Spores of food-poisoning strains of *B cereus* are much more heat-resistant than are spores of other strains from samples of uncooked rice. There are occasional reports of strains with even higher heat-resistance.[4] Since most strains of *B cereus* can form toxins, food-poisoning isolates and those from other sources are indistinguishable on this characteristic alone. Strains of *B cereus* that cause the diarrhoeal syndrome produce diarrhoeal enterotoxin, haemolysins, and lecithinase. The enterotoxin is a heat-labile protein with a molecular weight of about 38–46 kD. Strains associated with the emetic syndrome form a toxin that is probably a peptide with a low molecular weight (about 10 kD) which is highly stable; the toxin can withstand heating at 121°C for 90 min, and exposure to extremes of pH (2–11) and to the proteolytic enzymes trypsin and pepsin. It is possible that certain strains of *B cereus* produce either vomiting or diarrhoeal symptoms, or both, according to the nature of the food in which they have grown, but not all strains of *B cereus* can produce both types of toxin.

FOOD-POISONING DUE TO OTHER *BACILLUS* SPP

The general characteristics of *B subtilis* food-poisoning have been summarised as follows, largely on the basis of the UK incidents of food-poisoning (49 episodes, 175 cases from 1975 to 1986):[1] (a) foods most frequently involved are meat and pastry products and meat/seafood with rice dishes; (b) the illness has a short onset incubation period (median 2·5 h)—in almost a third of cases, onset times were 60 min or less; (c) vomiting, sometimes quite severe, is the predominant symptom in most cases—in just under half the cases this was followed by diarrhoea while about 10% of patients report headaches, flushing sensations, or sweating as additional symptoms; (d) large numbers (up to 10^7 cfu/g) of *B subtilis* have been found in the vomitus and the acute-phase faecal specimens of affected individuals; and (e) large numbers ($>10^5$–10^9) are needed to cause illness (Table 12.2). Peptide antibiotics produced by *B subtilis* and some other *Bacillus* spp during growth may inhibit competing bacteria and facilitate growth of *Bacillus*.

The major features of food-poisoning due to *B licheniformis* in outbreaks recorded in the UK (24 episodes, >218 cases, 1975–86) were that (a) the food vehicles most often involved were cooked meats and vegetables; (b) the median period of incubation was about 8 h and the predominant symptom was diarrhoea with vomiting in about half the cases (although the nausea, headaches, flushing, and sweating associated with *B subtilis* food-poisoning were not characteristic of *B licheniformis* food-poisoning); and (c) large numbers ($>10^6$ cfu/g) were found in the implicated foods and acute-phase faecal specimens. The main symptoms, incubation period, and food vehicles are similar to those of *Clostridium perfringens* food-poisoning (Table 12.2).

Only 5 incidents of food-poisoning associated with *B pumilus* were reported in the UK between 1975 and 1986.[1] These incidents involved meat dishes (2 incidents), scotch eggs, a cheese sandwich, and canned tomato juice. In each case gastroenteritis developed and large numbers of the bacterium, but no other established foodborne pathogens, were found in the food.

CONTROL MEASURES

Freshly cooked food eaten hot is safe, but if the food is allowed to cool slightly and is maintained at temperatures between 10°C and 60°C, spores that have survived cooking can germinate and the resulting vegetative bacteria can multiply in the food. The food either should be maintained at a temperature higher than 60°C or, if it is going to be stored, should be cooled rapidly to a temperature below 8–10°C to prevent growth or greatly reduce its rate.

Clostridium perfringens

OUTBREAKS OF FOODBORNE INFECTION

C perfringens, known as the cause of gas gangrene since the beginning of the century, also causes food-poisoning. The first report of foodborne illness due to this bacterium was in 1945 (McClung 1945, cited in ref 5) and it is clear that this disease is a world-wide concern. Strains of *C perfringens* are classified into five types, A-E, according to the extracellular toxins that are formed. Type A strains cause gas gangrene and are responsible for virtually all cases of foodborne illness and infectious diarrhoea due to *C perfringens* in human beings.

The number of reported outbreaks of food-poisoning due to *C perfringens* in England and Wales has fluctuated between 1975 and 1984, with 68 outbreaks and 1716 cases reported in 1984. Table 12.1 shows the number of cases from 1986 to 1988. Meat, meat products, and poultry were the suspected foods involved in most of the outbreaks. The food-poisoning is associated most commonly with food prepared in food-service establishments, including restaurants, institutions, hospitals, factories, schools, and caterers, most of whom need to prepare food well in advance of serving. Because the symptoms are mild, many outbreaks, especially those that result from domestic preparation of food, are probably not reported. Virtually no outbreaks have been linked to commercially prepared processed food.[2]

SYMPTOMS OF FOODBORNE INFECTION DUE TO TYPE A STRAINS

Symptoms usually occur 8–24 h after ingestion of heavily contaminated food—more than 10^5 vegetative bacteria/g.[5] The usual symptoms are diarrhoea and severe abdominal pain, less commonly with nausea, whereas fever and vomiting are unusual (Table 12.2.) Death is occasionally recorded in elderly and debilitated people. The occasional reports of sporadic illness, in which symptoms developed within 2 h, point to a role of preformed toxin.

The symptoms are due primarily to the enterotoxin, a protein of molecular weight of about 35 kD.[6] Sporulation of ingested bacteria is associated with the production of enterotoxin. The enterotoxin is not usually preformed in foods in high enough amounts to cause clinical illness. The binding of toxin to receptors on the surface of intestinal epithelial cells, and subsequent insertion into the cell membrane (McClane et al, 1988, cited in ref 5) leads to loss of structural integrity and function of the membrane and accumulation of fluid. The enterotoxin is destroyed at 60°C for 10 min.

NECROTIC ENTERITIS DUE TO TYPE C STRAINS

Very rarely, type C strains cause foodborne infection that results in necrotic enteritis. This disease was reported in Germany between 1946 and 1949 and was attributed to the consumption of underprocessed canned meat that contained heat-resistant type C spores.[6] It is now only reported in Papua New Guinea. The disease is due largely to the production by type C strains of the beta toxin, which is inactivated by proteolytic enzymes in the intestine. Susceptible people are those in whom production of these proteolytic enzymes is low because of their nutritional status or whose enzymes are impaired by inhibitors in the diet.

PROPERTIES OF C PERFRINGENS

The ability of C perfringens to multiply rapidly at a high temperature (optimum 43–45°C, range 15–50) is important in relation to food safety. The differences in reported heat-resistance of spores of C perfringens may be partly because detection of surviving spores on a culture medium may depend on the presence of lysozyme, which enables the germination of heat-damaged spores.[2]

Type A strains of C perfringens are usually present in soil at concentrations of 10^3–10^4/g. Types B, C, D, and E are obligate parasites, usually of domestic animals, and do not persist in soil. Type A strains occur widely in raw and processed foods but at numbers too low to cause infection. There are several reports that wild-type strains isolated from the environment and from foods do not form enterotoxin, whereas strains from cases of gastroenteritis do so. There is also a suggestion that exposure to heat renders the organism capable of forming enterotoxin, possibly because bacterial cells that are best able to sporulate are selected or because regulation of enterotoxin synthesis is disrupted.[6] The bacterium is part of the normal faecal flora of most individuals—numbers of spores range from 10^3 to 10^4/g. Patients in chronic care institutions may normally carry high numbers of spores of C perfringens in their intestinal tract and the detection of enterotoxin may be necessary to determine the cause of gastroenteritis.

CONTROL MEASURES

In virtually all outbreaks the principal cause is failure to refrigerate previously cooked foods properly, especially when foods have been cooked in large portions. Spores on raw foods, particularly meat and poultry, can survive cooking which will result in heat-activation and promote germination when the food reaches a suitable temperature during cooling. High numbers of vegetative bacteria may develop while the food is still palatable. It is impossible to prevent contamination of raw foods with the organism. To prevent *C perfringens* food-poisoning, cooked foods should be kept above 60°C or cooled to below 10°C within 2–3 h. An additional precaution is that cooked chilled foods should be heated to a minimum internal temperature of 75°C immediately before serving to destroy vegetative bacteria.

Clostridium botulinum

OUTBREAKS OF FOODBORNE BOTULISM

Growth of *C botulinum* in food or animal feed leads to the formation of a potent toxin. Ingestion of the toxin causes botulism, a neuroparalytic disease that affects man and animals and is often fatal unless treated promptly. Outbreaks and sporadic cases in human beings caused by growth of the bacterium in meat, fish, and vegetable foods have been reported world wide. The type of food involved has varied in different parts of the world, according to the dietary practices of the various communities.

Strains of *C botulinum* are classified into several types (A–G) based on the antigenic properties of the toxin formed. Types A, B, E, and F are responsible for most of the cases of human botulism, whereas types C and D usually cause botulism in animals and birds. Strains that cause botulism in man are divided into two groups, proteolytic and non-proteolytic, according to their ability to hydrolyse certain proteins. All type A strains are proteolytic and type E strains are almost invariably non-proteolytic. Types B and F are divided into proteolytic and non-proteolytic strains. Proteolytic strains of types A, B, and F form spores with a relatively high heat-resistance, a sodium chloride concentration of 10% (weight/volume) is required to inhibit growth, and the bacteria will not multiply at a temperature of 10°C or lower. Non-proteolytic strains of types B, E, and F form spores that are less heat-resistant, growth of the bacteria is inhibited by a sodium chloride concentration of 5% (weight/volume), and the minimum temperature for growth is about 3·3°C (Table 12.2).

C botulinum was first isolated in 1895 after an outbreak of botulism in Belgium. From the 1950s to the 1980s the foods most frequently implicated in botulism in the USA (apart from Alaska), Spain, Italy, and China were vegetables, particularly after home-preservation by heat treatment or by fermentation.[7] In Alaska, Canada, Japan, and Scandinavia most outbreaks have been associated with fish products, whereas in Germany, Italy, France, and Poland the main foods involved have been meats—eg, home-cured hams. Since 1950, most cases of botulism have been associated with foods prepared in the home or in small catering units rather than with commercially prepared foods. Data about the incidence of botulism in Europe are difficult to obtain; however, between 1979 and 1988, 148 outbreaks (>225 cases and 4 deaths) were recorded at the Institut Pasteur, France, and it is estimated that these represented about half the total number of outbreaks.

In the UK the incidence of foodborne botulism is much lower than that in the USA or Europe. Between 1922 and 1988, 9 outbreaks (26 cases) were reported in Britain; most of them involved home-prepared food but the largest, which occurred in 1922 and affected 8 people, all of whom died, was due to a commercially prepared duck paste. In 1989, the largest recorded outbreak of foodborne botulism in the UK affected 27 people, of whom 26 were admitted to hospital, 12 were treated in intensive care units, 8 received positive ventilation, and 1 died.[8] The disease was associated with the consumption of a hazelnut yoghurt. The yoghurt was too acid to allow growth of C botulinum and the botulism was caused by type B toxin formed by growth of the bacterium in the canned hazelnut conserve, which had been added to the yoghurt and which had been inadequately heat treated.[8]

SYMPTOMS OF BOTULISM

C botulinum toxin is lethal; the ingestion of as little as 0·1 g of a food in which the bacterium has grown can cause botulism.[9] The incubation period is usually 12–36 h but can be between 2 h and 8 days.[3] Symptoms include disturbances of vision (including diplopia), dysphagia, generalised weakness of the limbs and respiratory muscles, and often nausea and vomiting. Botulism has been confused with other diseases, including the Guillain-Barré syndrome and myasthenia gravis, and with the effects of other food-poisoning bacteria.

INFANT BOTULISM

Infant botulism was first recognised in the USA in 1976 where it is now the most common form of botulism[5] (as many as 99 cases have been reported in one year). There is evidence that in certain infants under a year old, ingested spores of the bacterium can germinate and multiply in the intestinal tract before the establishment of the normal adult gut flora that would inhibit growth of C botulinum. Infant botulism has been reported from North and South America, Europe, and Australia; 2 cases have been reported in England. In 1979, a strain of C barati that produced type F botulinum toxin was isolated from a case of infant botulism in New Mexico. The first 2 cases of infant botulism reported in Italy, in 1986, were remarkable because they were the first cases found to be caused by type E botulinum toxin and because the organism isolated that formed the toxin was identified not as C botulinum but as C butyricum. Despite the fact that type C and D strains of C botulinum rarely cause botulism in man, a case of infant botulism due to a type C strain has been reported recently from Japan.[14]

PROPERTIES OF C BOTULINUM

C botulinum is present in soil and the environment but to a lesser extent compared with C perfringens. The prevalence of C botulinum in foods is generally low. The spores probably occur most frequently in fish products (1–29%), the number of spores varying from 1 to 170/kg.[7] Clearly, much lower levels of contamination have important implications in food production. If growth of the organism is prevented, spores are usually of little importance except in infant foods. Honey has been implicated as the probable source of C botulinum spores in at least 20 cases of infant botulism in California; 80 spores/g were reported in one sample of incriminated honey.[10] The high sugar content of honey will prevent the growth of *C botulinum*, and contamination may result from multiplication of the organism in dead larvae in the hives. In a survey in the UK, spores could not be detected in 20 g portions of 122 samples of honey.[11]

C botulinum requires anaerobic conditions for growth. The fact that in some foods the nature of the food and the packaging may result in too high a concentration of oxygen to allow growth cannot always be relied upon. Some recent outbreaks in the USA have resulted from growth of the bacterium in foods that might not have been expected to be anaerobic—eg, in bottled garlic in oil, baked potatoes wrapped in aluminium foil, and sautéed onions.

PROPERTIES OF *C BOTULINUM* TOXINS

Most strains of *C botulinum* produce toxin of a single antigenic type. Very rarely, strains have been isolated that form both the major type-specific toxin and a smaller amount of toxin of another type. The toxins are proteins and may exist in four molecular sizes (150–900 kD).[7] The larger forms ("progenitor toxins") are complexes of the smaller 150 kD molecular weight toxic component (the "derivative toxin" or the neurotoxin) with non-toxic components. The lethality of the toxins, especially those produced by non-proteolytic strains, may be increased substantially by activation with trypsin.

The oral lethal dose for man is between 0·005 and 0·1 μg for toxin of proteolytic strains and between 0·1 and 0·5 μg for toxin of non-proteolytic strains (calculated from ref 7). Botulinum toxins are neurotoxic—ie, they block the release of transmitter at neuromuscular junctions. Both cholinergic and adrenergic systems are affected but much higher concentrations of toxin are required to inhibit the release of noradrenaline than of acetylcholine.

The stability of the toxins during heating depends on the nature of the heating medium; proteins, colloids, and high ionic strength have a protective effect. To inactivate any botulinum toxin at concentrations that can be formed in foods, time and temperature combinations of, for example, 20 min at 79°C or 5 min at 85°C have been recommended.[7]

The stability of botulinum toxin to irradiation also depends on the medium in which it is present. Food components can be highly protective and in a meat slurry an irradiation dose of 10 kGy (which is the overall average dose that has been accepted as suitable for use in food-preservation[12] and proposed for use in the UK) was not sufficient to destroy high levels of toxin.[13] A code of Good Manufacturing Practice, similar to that currently applied in the heat processing of food, would be used to ensure that growth of *C botulinum* and formation of its toxin was prevented in the food before irradiation.

CONTROL MEASURES

In general, *C botulinum* cannot grow in high-moisture acidic foods (pH below 4·5), although the spores can remain viable. The preservation of high-moisture foods with a pH of 4·5 or higher (low-acid foods) is geared particularly to the control of *C botulinum*; very often this also ensures the control of other foodborne pathogens and of spoilage microorganisms.

Canned low-acid foods are preserved by a heat process that destroys the spores of *C botulinum*, whereas most other high-moisture low-acid foods are preserved by a combination of factors that inhibit their germination and growth. These factors include mild heat treatment, addition of salt and nitrite, and refrigeration. Whereas the canning process is designed to reduce the number of surviving spores of proteolytic strains by 10^{12}-fold, the extent by which combinations of inhibitory factors reduce the risk of growth of *C botulinum* is often less well-defined. With trends to decrease the content of salt, nitrite, and fat in foods; to increase the production of foods preserved by a mild heat treatment followed by refrigeration; and to extend the use of vacuum packaging there is an increased reliance on refrigeration for the safety of foods. Because non-proteolytic strains can grow at temperatures as low as $3 \cdot 3°C$, the ability of heat processing and of other inhibitory factors to control these strains is currently of concern to the food industry, and research is in progress to improve the effectiveness of this control. A high margin of safety against growth and toxin production by *C botulinum* must be built into the production of processed foods. The major assessments on which such safety margins are calculated are (a) the extent of lethality or inhibition of *C botulinum* provided by the processing, the preservation system, or both; and (b) the estimated frequency of occurrence of the bacterium in the food or the food environment. Generally, very low numbers of *C botulinum* spores in a food that will not allow growth of the bacterium are not important, but in food for infants or for people undergoing gastrointestinal surgery or being treated with antibiotics there is a low risk of botulism.

A high proportion of the cases of botulism world wide results from preparation of food on the domestic scale and in small catering units. In the UK, the home-canning of vegetables is discouraged because of the risk of survival and growth of the spores. It is important that those who devise and publish new methods for the preservation and preparation of food in the home and in catering units should ensure that the processes provide an acceptable degree of safety against the risk of growth of this bacterium.

I thank Dr M. O'Mahoney, Dr E. Mitchell, and Mr Paul Sockett of the PHLS Communicable Disease Surveillance Centre, London, UK, for supplying information and Mr J. M. Kramer of the Central Public Health Laboratory for helpful discussions.

References

1. Kramer JM, Gilbert RJ. *Bacillus cereus* and other *Bacillus* species. In: Doyle MP, ed. Foodborne bacterial pathogens. New York: Marcel Dekker, 1989: 21–70.
2. Labbe R. *Clostridium perfringens*. In: Doyle MP, ed. Foodborne bacterial pathogens. New York: Marcel Dekker, 1989: 191–234.
3. Smith LDS. Botulism. The organism, its toxins, the disease. Illinois: Thomas, 1977.
4. Bradshaw JG, Peeler JT, Twedt RM. Heat resistance of ileal loop reactive *Bacillus cereus* strains isolated from commercially canned food. *Appl Microbiol* 1975; **30**: 943–45.
5. Hatheway CL. Toxigenic clostridia. *Clin Microbiol Rev* 1990; **3**: 66–98.
6. Granum PE. *Clostridium perfringens* toxins involved in food poisoning. *Int J Food Microbiol* 1990 ; **10**: 101–12.

7. Hauschild AHW. *Clostridium botulinum*. In: Doyle MP, ed. Foodborne bacterial pathogens. New York: Marcel Dekker, 1989: 111–89.
8. O'Mahoney MO, Mitchell E, Gilbert RJ, et al. An outbreak of foodborne botulism associated with contaminated hazelnut yoghurt. *Epidemiol Infect* 1990; **104:** 389–95.
9. Lamanna C, Carr CJ. The botulinal, tetanal and enterostaphylococcal toxins. *Clin Pharmacol Ther* 1967; **8:** 286–332.
10. Midura TF, Snowden S, Wood RM, Arnon SS. Isolation of *Clostridium botulinum* from honey. *J Clin Microbiol* 1979; **9:** 282–83.
11. Berry PR, Gilbert RJ, Oliver RWA, Gibson AAM. Some preliminary studies on the low incidence of infant botulism in the United Kingdom. *J Clin Pathol* 1987; **40:** 121.
12. Report on the safety and wholesomeness of irradiated foods. Advisory Committee on Irradiated and Novel Foods. London: Department of Health, 1986.
13. Rose SA, Modi NK, Tranter HS, Bailey NE, Stringer MF, Hambleton P. Studies on the irradiation of toxins of *Clostridium botulinum* and *Staphylococcus aureus*. *J Appl Bacteriol* 1988; **65:** 223–29.
14. Oguma K, Yokota K, Hayashi S, et al. Infant botulism due to *Clostridium botulinum* type C toxin. *Lancet* 1990; **336:** 1449–50.

13
Foodborne staphylococcal illness

Howard S. Tranter

Staphylococcus aureus causes food-poisoning by the production of one or more heat-stable extracellular toxins, which are wholly responsible for the symptoms of the disease.[1] The syndrome is characterised by nausea, vomiting, abdominal pain, and diarrhoea, 2–6 h after eating contaminated food. Although prostration can occur in severe cases, recovery usually takes 1–3 days and death is rare. Time to onset and severity of symptoms depend on the amount of toxin consumed and an individuals' susceptibility to the toxin. Staphylococcal food-poisoning is not a reportable disease in many countries, so its true incidence is unknown. Most cases probably go unrecognised because of their short duration; only outbreaks which involve large numbers of people (eg, at picnics, group dinners, or public institutions) come to the attention of the public health authorities. The incidence of the disease in different countries varies according to geography and eating habits. For example, in the USA from 1983 to 1987, 47 (7·8%) of the recorded 600 bacterial food-poisoning outbreaks were due to *S aureus*,[2] whereas during the same period in England and Wales *S aureus* accounted for 1·9% of a total of 2815 bacterial food-poisoning (54/2815) outbreaks.

Epidemiology

Foods most often incriminated in staphylococcal foodborne disease include cooked meat, fish, poultry, bakery foods (especially those with cream or custard fillings), dairy produce, fruit, vegetables, and salads. Poor handling of foods in food service establishments seems to be the major cause of outbreaks. For example, about a third of the staphylococcal outbreaks (7/26) recorded in England and Wales from 1986 to 1988 were associated with retail premises. Mishandling of food in the home is also important. However, few outbreaks can be traced directly to contamination during food processing. Most outbreaks are due to contamination of foods by foodhandlers. 20–50% of healthy individuals carry *S aureus*; the nose is the main site of multiplication, and staphylococci are also found on the skin and in faeces. Therefore, education in the hygienic preparation of food is essential. *S aureus* may be present in raw milk from mastitic cows; pasteurised milk is regarded as safe, although in a survey in Brazil, counts of *S aureus* as high as 1000/ml were recorded.[3] There is also a potential for cross-contamination of cooked food by raw food, especially by meat. The use of raw materials of high microbiological quality and their maintenance in a state that will prevent staphylococcal growth is important because even if the end product is subsequently heat-processed to destroy viable staphylococci the heat-stable enterotoxins may be unaffected. Food in which staphylococci survive but cannot grow may be hazardous if the food is used as an ingredient in other foods in which the staphylococci can grow. Products that are semi-preserved with salt or sugar (eg, cooked meats) may favour the growth of staphylococci; unlike many other organisms, staphylococci can tolerate these compounds. Cooked high-protein foods, in which staphylococci face no competition from other organisms, present high risks.

To prevent staphylococcal food-poisoning, it is important to keep susceptible foods refrigerated at all times except during preparation and serving. Rapid cooling of the food is especially important; most food-poisoning outbreaks could be prevented if this simple precaution were taken.

Although 50–70% of *S aureus* strains are enterotoxigenic, tests for toxin production alone may not be sufficient to identify strains responsible for an outbreak. Phage typing[4] has been used successfully to characterise the organism for more than 35 years and has proved useful in epidemiological studies. Most strains implicated in food-poisoning outbreaks are lysed by phages of group III or groups I and III, although many enterotoxigenic strains are not susceptible to any phages and are therefore non-typable. Other coagulase-positive staphylococci—namely, *S hyicus* and *S intermedius*—produce small amounts of enterotoxin but these organisms have not yet been implicated in outbreaks of food-poisoning.

Staphylococcal enterotoxins

The staphylococcal enterotoxins (SE) are a group of seven (SEA, SEB, SEC_1, SEC_2, SEC_3, SED, and SEE) serologically distinct exoproteins, the production of which depends on various environmental factors.[5] Many foods will support growth of *S aureus* and toxin formation with the exception of those of lower pH ($<5 \cdot 0$) or a water activity (a_w) below $0 \cdot 86$. The toxins are produced over a wide range of temperatures (10–45°C; optimum 35–40°C). Growth of *S aureus* and toxin production may be restricted by other competing microorganisms—for example, spoilage organisms in raw or fermented foods. However, foods with low a_w, such as hams, cured meats, and sweet cream products, do not support the growth of these spoilage bacteria, and the staphylococci can grow unchecked, unless storage temperature prevents growth. Enterotoxin serotypes A, D, or A with D more frequently cause food-poisoning possibly because these toxins can be produced over a wider range of growth conditions or in the presence of fewer staphylococci than the other serotypes.

The purified toxins are small (MW range $27 \cdot 1$–30 kD) single-chain polypeptides with similar basic structures; each has a single disulphide loop near the centre of the molecule. Microheterogeneity, as shown by differences in isoelectric points (Table 13.1), has been observed in purified toxin preparations and may result from deamidation of the toxins either naturally or enzymically during fermentation or purification processes. There are 2 groups of enterotoxins, according to whether they share antigenic sites—namely, SEB and the SECs; and SEA, SED, and SEE. In all the toxin types there is only one region in which there is a common sequence of more than 4 aminoacid residues that includes the only conserved histidine residue. Chemical modification of these histidine residues in either SEA[6] or SEB[7] leads to loss of emetic activity but little change in molecular conformation of the toxins; this supports the hypothesis that the conserved histidine domain may be associated with a toxic active site.

PHYSICOCHEMICAL PROPERTIES

The staphylococcal enterotoxins are believed to be thermostable molecules inactivated only by prolonged boiling. The amount of inactivation depends not only on the temperature but also on the composition and pH of the medium and the purity of the toxin preparations. Some studies[8] suggest that loss of immunological activity is greater at 70–80°C than at 90–100°C. Tatini[9] also reported that there was an increase in biological activity of the toxins after heat treatment. Lately, Schwabe et al[10] found that toxin which had been completely inactivated by the heating of toxin-containing food extracts could regain both biological and immunological activity after the extracts were adjusted to pH 11·0 followed by readjustment to pH 7·0. All the toxins seem to be stable at high pH (up to pH 11) but less so at lower pH (down to pH 2·5).

Resistance of the enterotoxins to proteolysis by enzymes such as trypsin, chymotrypsin, pepsin, and papain, probably allows the toxins to pass, without loss of activity, through the stomach to the intestinal tract, where they stimulate emesis and diarrhoea. However, SEB and SEC_1 are readily nicked by limited digestion with trypsin to form 2 and 5 fragments, respectively.

Table 13.1 Characteristics of the staphylococcal enterotoxins

Entero-toxin serotype	Molecu-lar weight (D)*	pI value of major forms	pH range of isoelectric forms	Toxin gene location
A	27 100	8·1; 7·3	6·5–8·6	Chromosomal
B	28 366	8·3; 8·6 9·1; 9·4	7·8–9·6	Chromosomal/plasmid
C_1	27 496	8·8; 9·2	7·9–9·2	Chromosomal/plasmid
C_2	27 531	5·6; 6·9 7·5; 7·6	5·5–9·2	NK
C_3	27 438	8·2	6·6–8·3	NK
D	26 360	8·4	7·9–9·3	Plasmid
E	26 425	8·0	7·0–8·5	Chromosomal

NK = not known.
*Calculated from nucleotide sequence.

The biological activity of the toxins has been primarily associated with the larger C-terminal peptides of SEB (MW 17 kD) and SEC_1 (MW 19 kD). A small N-terminal fragment of SEC_1 (MW 6782 D) has been purified and reported to have mitogenic activity,[11] though it cannot be ruled out that this activity was due to low level contamination by intact toxin molecules.

DETECTION

Although contaminated foods frequently contain large numbers (1×10^6/g) of enterotoxigenic staphylococci, use of biochemical tests (eg, those for coagulase or thermonuclease activity) for diagnosis are inappropriate if the contaminating staphylococci die during storage or are destroyed by food processing. Furthermore, enterotoxin production by coagulase-negative and thermonuclease-negative staphylococci is not unknown. On the other hand, since foodborne toxins are resistant to high temperatures and irradiation, confirmation of outbreaks can be made by direct detection of toxin in suspect food. Because there are no direct chemical tests, staphylococcal enterotoxins are usually detected by immunologically-based assays.[12]

The amount of enterotoxin required to cause illness in man is not known, but at least 1 μg toxin/100 g of food will induce clinical symptoms; thus, a diagnostic test must be able to detect 1 ng toxin/g of food. Detection systems based on the formation of an antibody-antigen precipitate in agar/agarose gels, such as the microslide test, are still used for the detection of staphylococcal enterotoxins but require lengthy extraction and concentration procedures to isolate the toxin from specimens for adequate sensitivity. More sensitive techniques with labelled antibodies, such as radioimmunoasay and enzyme-linked immunosorbent assay,[13] can detect as little as 0·1–1·0 ng toxin/g of food. A reversed passive latex agglutination (RPLA) test for the detection of enterotoxins A, B, C, and D (Oxoid, Basingstoke, UK) uses latex particles that have been sensitised with purified anti-enterotoxin immunoglobulins; these agglutinate in the presence of homologous toxin. The sensitivity of the assay is 0·5 ng/ml, which is adequate to detect levels of toxin implicated in most, if not all, food-poisoning outbreaks. Although the RPLA is easy to perform and requires no special equipment, the test is subjective, requires 24 h to complete, and occasionally gives non-specific agglutinations with some foods.[14]

BIOLOGICAL ACTIVITIES

Although immunological assays are useful for detecting toxin in food samples, such assays are presently unable to distinguish between biologically active and inactive toxin; biological activity can only be detected in animals. Most animals, apart from kittens and monkeys, are moderately insensitive to the enterotoxins. All staphylococcal enterotoxins cause emesis in monkeys after intragastric or intravenous administration with an emetic dose per animal of 5–10 μg and 20–500 ng, respectively. Although there are numerous descriptive studies on the physiological effects of the staphylococcal enterotoxins, the molecular basis behind such effects remains unclear. The emetic response may be due to stimulation of nerve centres in the gut, which is transmitted to the vomiting centre in the brain via the vagus and sympathetic nerves. If these two nerves are severed, vomiting does not occur. The diarrhoeal response to the enterotoxins is even less well understood; toxin passes through the rat intestinal wall into the blood circulation before elimination by the kidneys.[15] Unlike the enterotoxins of *Escherichia coli* and *Vibrio cholerae*, the staphylococcal enterotoxins do not exert their effect by stimulation of adenylate cyclase activity.

Lately, the role of the staphylococcal enterotoxins as "superantigens", which cause activation and proliferation of T lymphocytes, has been recognised.[16] It is now apparent that the staphylococcal enterotoxins can stimulate production of various chemical messengers, including interferon, interleukin-1, and tumour necrosis factor. In this respect, the enterotoxins resemble another toxin produced by *S aureus*—namely, toxic shock syndrome toxin-1 (TSST-1)—and this may explain the recent observations[17] that SEA, SEB, and SEC produced by clinical strains of *S aureus* can induce toxic-shock-like illness. These new observations may well contribute to our understanding of the biochemical nature of the toxins action. For example, it is possible that T-cell activation accompanied by secretion of lymphokines is responsible wholly or partly for some of the pathological effects caused by these toxins in man, including vomiting, diarrhoea, and shock.

I thank Dr P. Hambleton for constructive comments during the preparation of this manuscript.

References

1. Bergdoll MS. *Staphylococcus aureus*. In: Doyle MP, ed. Bacterial foodborne pathogens. New York: Marcel Dekker, 1989: 464–523.
2. Bean NH, Griffin PM, Goulding JS, Ivey CB. Foodborne disease outbreaks, 5-year summary, 1983–1987. *MMWR* 1990; **39**: 15–57.
3. Santos EC dos, Genigeorgis C, Farver TB. Prevalence of *Staphylococcus aureus* in raw and pasteurized milk used for commercial manufacturing of Brazilian Minas Cheese. *J Food Protect* 1981; **44**: 172–76.
4. De Saxe M, Coe AW, Wieneke AA. The use of phage typing in the investigation of food poisoning caused by *Staphylococcus aureus* enterotoxins. *Soc Appl Bacteriol Tech Ser* 1982; **17**: 173–97.
5. Halpin–Dohnalek MI, Marth EH. *Staphylococcus aureus:* production of extracellular compounds and behaviour in foods—a review. *J Food Protect* 1989; **52**: 267–82.
6. Stelma GN, Bergdoll MS. Inactivation of staphylococcal enterotoxin A by chemical modification. *Biochem Biophys Res Commun* 1982; **105**: 121–26.
7. Scheuber PH, Golecki JR, Kickhofen F, Scheel D, Beck G, Hammer DK. Skin reactivity of unsensitised monkeys upon challenge with staphylococcal enterotoxin B: a new approach for investigating the site of toxic action. *Infect Immun* 1985; **50**: 869–76.
8. Modi NK, Rose SA, Tranter HS. The effects of irradiation and temperature on the immunological activity of staphylococcal enterotoxin A. *Int J Food Microbiol* 1990; **11**: 85–92.
9. Tatini SR. Thermal stability of enterotoxins in food. *J Milk Food Technol* 1976; **39**: 432–38.
10. Schwabe M, Notermans S, Boot R, Tatini SR, Kramer J. Inactivation of staphylococcal enterotoxins by heat and reactivation by high pH treatment. *Int J Food Microbiol* 1990; **10**: 33–42.

11. Spero L, Morlock BA. Biological activities of the peptides of staphylococcal enterotoxin C formed by limited tryptic hydrolysis. *J Biol Chem* 1978; **253**: 8787–91.
12. Tranter HS, Brehm RD. Production, purification and identification of the staphylococcal enterotoxins. *Soc Appl Bacteriol Symp Ser* 1990; **19**: 109S–122S.
13. Fey H. Staphylococcal enterotoxins. In: Kohler RB, ed. Antigen to diagnose bacterial infection. Vol 2. Florida: CRC Press, 1986: 211–38.
14. Berry PR, Rodhouse JC, Wieneke AA, Gilbert RJ. Use of commercial kits for the detection of *Clostridium perfringens* and *Staphylococcus aureus* enterotoxins. *Soc Appl Bacteriol Tech Ser* 1987; **24**: 245–54.
15. Beery JT, Taylor SL, Schlunz LR, Freed RC, Bergdoll MS. Effects of staphylococcal enterotoxin A on the rat gastrointestinal tract. *Infect Immun* 1984; **44**: 234–40.
16. Marrack P, Kappler J. The staphylococcal enterotoxins and their relatives. *Science* 1990; **248**: 705–11.
17. Yaqoob M, McClelland P, Murray AE, Mostafa SM, Ahmad R. Staphylococcal enterotoxins A and C causing toxic shock syndrome. *J Infect* 1990; **20**: 176–77.

14
Foodborne viruses

Hazel Appleton

Unlike bacteria, viruses do not multiply or produce toxins in food; food items merely act as vehicles for their transfer. There is the potential for any enteric virus which causes illness when ingested to be transmitted by food, but in practice most reported incidents of viral foodborne illness are due to gastroenteritis viruses and hepatitis A virus (HAV). Although such viruses are usually transmitted from person to person, it is increasingly recognised that food may have an important role in their transmission. It is likely that foodborne transmission of viral infections is greatly under-reported, and the extent of the problem is not known.

Epidemiology

Foods may be contaminated either at source (primary contamination) or at the time and place of preparation (secondary contamination).

PRIMARY CONTAMINATION

The most clearly implicated food in the transmission of viruses is the bivalve mollusc (oysters, clams, cockles, and mussels); most illness associated with consumption of this type of shellfish is viral (Table 14.1). Some outbreaks in various parts of the world have involved several hundred or more people (refs 1–5, and Public Health Laboratory Service [PHLS] Communicable Disease Surveillance Centre, unpublished). Bivalve molluscs are filter feeders that inhabit shallow inshore coastal waters, which may be polluted with sewage. During their natural feeding process, molluscs extract particulate matter, including bacteria and viruses, from the very large volumes of water that pass over their gills. These molluscs are often not cooked thoroughly but may be subject to a brief heat treatment only and oysters are frequently eaten raw. Oysters are removed from their growing waters before

Table 14.1   Outbreaks of illness associated with molluscs: England and Wales, 1965–88

Type of outbreak	No of outbreaks
Bacterial food-poisoning	12
Hepatitis A	17
Viral gastroenteritis	37
Paralytic shellfish poisoning	1
Red whelk poisoning	1
Unknown*	101
Total	*169*

*Features often characteristic of viral gastroenteritis.
Compiled from published and unpublished data from PHLS Communicable Disease Surveillance Centre, Food Hygiene Laboratory, and Virus Reference Laboratory.

they are marketed, and placed in clean water, usually in purpose-built tanks supplied with ultraviolet-irradiated or ozone-treated water. As a result of this process, bacteria are eliminated within 24 h, but viruses are not always removed. Work is urgently needed to establish conditions for the removal of viruses since there is no assurance yet that oysters may be eaten with safety.

Other molluscs, such as cockles, may be heat treated before sale but prolonged cooking can lead to an unpalatable product. Studies on the heat inactivation of HAV[6] have led to recommendations to the food industry that the internal temperature of shellfish meat should be raised to 85–90°C and maintained for 1·5 min. This regimen will kill HAV, but since most gastroenteritis viruses cannot be cultured in vitro, there is no way to confirm that these viruses will be inactivated. Nonetheless, since early 1988, when recommendations from the Ministry of Agriculture, Fisheries and Food were implemented in the UK industry, there have been no reported outbreaks from England and Wales of viral illness from this source, although shellfish that have not been heat treated have continued to be associated with illness (P. Sockett, personal communication). However, the past two winters in the UK have been exceptionally mild, and it is noteworthy that viral contamination has often followed very cold periods when heat treatment procedures may have been inadequate.

Another potential source of primary viral contamination is the application of polluted water and sewage sludge to fruit and vegetable crops. Viruses may be transferred by direct surface contact during irrigation and fertilisation, and it has also been suggested from experimental studies that mammalian viruses could be taken up through the roots from ground waters.[7] Salad items are frequently implicated in outbreaks, though contamination is usually believed to have occurred during preparation.[8,9]

SECONDARY CONTAMINATION

Secondary contamination arises from infected foodhandlers, and is largely associated with cold food items that require much handling during preparation—eg, sandwiches and salads; surfaces in food preparation areas may become contaminated. The viruses involved are highly infectious in low doses. There is circumstantial evidence that some gastroenteritis outbreaks may be caused by symptom-free excretors of viruses,[10] but without conclusive evidence for such carrier states there is no reason why foodhandlers should be excluded from work for more than 48 h after symptoms have stopped. For hepatitis A, the main infectious period precedes symptoms and so exclusion of infectious personnel is not possible. Control depends on scrupulous attention to good hygienic practice, including frequent hand washing and careful separation of raw and cooked foods; molluscs should be regarded in the same way as uncooked meat. For disinfection of contaminated surfaces, chlorine-based compounds are probably the most appropriate.

Hepatitis viruses

Waterborne and foodborne spread of hepatitis A has long been recognised, though in the developed world waterborne outbreaks are rare and foodborne outbreaks are uncommonly reported. It is difficult to associate illness with a specific food item because of the long incubation period of hepatitis A (3–6 weeks); for the same reason, foodborne transmission is probably under-recognised. Although many food items have been implicated, shellfish account for a high proportion of outbreaks.[9,11] Soft fruits, such as raspberries and strawberries, either raw or processed into ice cream and other foods have also been implicated.[12] The incidence of hepatitis A in the UK has been rising since 1987. It may be reasonable to assume that increased prevalence of circulating virus and discharge in sewage could result in increased contamination of shellfish.

The recently identified enteric form of hepatitis (designated hepatitis E) has been associated with large waterborne outbreaks in some developing countries, and there is a potential for foodborne transmission of this virus, particularly by shellfish.

Gastroenteritis viruses

Viruses of several morphological groups have been associated with gastroenteritis. However, more than 90% of foodborne outbreaks in which a virus has been identified have been attributed to one of the small round structured viruses (SRSV) of the Norwalk group;[9] caliciviruses, astroviruses, parvoviruses, and rotaviruses have only rarely been implicated. When sewage pollution is involved, as in shellfish contamination, more than one virus type may be detected.[2,13] Instances of gastroenteritis 24 h after eating shellfish, followed by hepatitis A 3–4 weeks later, have been recorded.

Viral gastroenteritis has a variable incubation period (usually 15–50 h), which may be dose dependent, and symptoms often include both vomiting and diarrhoea. Secondary cases (person-to-person transmission) are a characteristic feature. Reports of foodborne outbreaks of viral gastroenteritis in the UK have increased in recent years possibly as a result of increased surveillance and awareness. However, because the illness is mild, most cases of viral gastroenteritis remain unreported; hence the true scale of the problem is unknown. Although symptoms may be mild and short-lived, large foodborne outbreaks may have considerable economic impact in terms of working days lost.[9] They may disrupt or even lead to the closure of hospitals and schools (refs 14, 15, and PHLS Communicable Disease Surveillance Centre, unpublished).

Virus detection

Detection of gastroenteritis viruses largely depends on examination of faecal specimens by electronmicroscopy. The number of SRSVs rapidly falls below detectable levels and so specimens should ideally be collected within 48 h of onset of symptoms. Freezing and thawing may sometimes destroy the morphological integrity of virions but not the infectivity of the sample. Specimens should therefore be stored without freezing before examination. It is not practicable to examine food remnants because the number of virus particles will be much too low for currently available detection methods. The general view is that at least 10^6 virus particles/ml of specimen are required for detection by electronmicroscopy. Viruses with very distinct morphological features—eg, rotavirus—will be more readily detected than the small round viruses whose features are less clearly defined, and it may be necessary for greater numbers of these viruses to be present for definitive identification.

Since HAV is excreted mainly before infection is apparent, faecal specimens are inappropriate. Laboratory diagnosis of infection depends on detection of specific anti-HAV IgM in serum. HAV can be grown in cell culture and this provides the opportunity to investigate some foodborne outbreaks more thoroughly. However, primary virus isolation is a lengthy procedure and isolation from a food source has never been achieved. Recently HAV was isolated from a drinking water supply that had been responsible for an outbreak in the USA, but this involved culture of highly concentrated water samples for up to 21 weeks.[16] Specific gene probes to detect HAV in faecal samples[17] have been developed but these have not been successfully applied to food and environmental specimens. Further work will be needed to enhance their sensitivity.

The attempts by several investigators to develop gene probes for detection of SRSV are hampered by the failure to grow the virus in vitro and the difficulty in extracting virus directly from clinical specimens. Despite these difficulties, new developments in molecular biology techniques offer the most promising approach to detection of virus in food and water and in associated illnesses.

References

1. Appleton H, Pereira MS. A possible virus aetiology in outbreaks of food poisoning from cockles. *Lancet* 1977; i: 780–81.
2. Murphy AM, Grohmann GS, Christopher PJ, Lopez WA, Davey GR, Millsom RH. An Australia-wide outbreak of gastroenteritis from oysters caused by Norwalk virus. *Med J Aust* 1979; 2: 329–33.
3. O'Mahony MC, Gooch CD, Smyth DA, Thrussell AJ, Bartlett CLR, Noah ND. Epidemic hepatitis A from cockles. *Lancet* 1983; i: 518–20.

4. Gill ON, Cubitt WD, McSwiggan DA, Watney BM, Bartlett CLR. Epidemic of gastro-enteritis caused by oysters contaminated with small round structured viruses. *Br Med J* 1983; **287**: 1532–34.
5. World Health Organisation. Outbreak of hepatitis A: Shanghai. *Wkly Epidemiol Rec* 1988; **63**: 91–92.
6. Millard J, Appleton H, Parry JV. Studies on heat inactivation of hepatitis A virus with special reference to shellfish. *Epidemiol Infect* 1987; **98**: 397–414.
7. Katzenelson E, Mills D. Contamination of vegetables with animal viruses via the roots. *Monogr Virol* 1984; **15**: 216–20.
8. Griffin MT, Surowiec JL, McCloskey DI, et al. Foodborne Norwalk virus. *Am J Epidemiol* 1982; **115**: 178–84.
9. PHLS Working Party on Viral Gastroenteritis. Foodborne viral gastroenteritis (with a brief comment on hepatitis A). *PHLS Microbiol Dig* 1988; **5**: 69–75.
10. Iverson AM, Gill M, Bartlett CL, Cubitt WD, McSwiggan DA. Two outbreaks of foodborne gastroenteritis caused by a small round structured virus: evidence of prolonged infectivity in a food handler. *Lancet* 1987; ii: 556–58.
11. Cliver DO. Vehicular transmission of hepatitis A. *Public Health Rev* 1985; **13**: 235–92.
12. Reid TMS, Robinson HG. Frozen raspberries and hepatitis A. *Epidemiol Infect* 1987; **98**: 109–12.
13. Appleton H. Small round viruses: classification and role in foodborne infections. In: Bock J, Whelan J, eds. Novel diarrhoea viruses. (CIBA Foundation Symposium 128). Chichester: Wiley, 1987: 108–25.
14. Kuritsky JN, Osterholm MT, Greenberg HB, et al. Norwalk gastroenteritis: a community outbreak associated with bakery product consumption. *Ann Intern Med* 1984; **100**: 519–21.
15. Pether JVS, Caul EO. An outbreak of foodborne gastroenteritis in two hospitals associated with a Norwalk-like virus. *J Hyg (Camb)* 1983; **91**: 343–50.
16. Bloch AB, Stramer SL, Smith JD, et al. Recovery of hepatitis A virus from a water supply responsbile for a common source outbreak of hepatitis A. *Am J Public Health* 1990; **80**: 428–30.
17. Ticehurst JR, Feinstone SM, Chestnut T, Tassopoulos NC, Popper H, Purcell RH. Detection of hepatitis A virus by extraction of viral RNA and molecular hybridization. *J Clin Microbiol* 1987; **25**: 1822–29.

15
Foodborne protozoal infection

David P. Casemore

Foodborne transmission of protozoa seems to be uncommon. This is surprising in view of the cosmopolitan prevalence and the simple faecal-oral mode of transmission of some species. Infection in man is often symptomless, but may produce overt disease (particularly gastroenteritis or other enteropathy), and in some cases causes more generalised systemic disease. Protozoa are single-celled microorganisms with simple or complex life cycles. Many can encyst, which provides the means for survival in the environment and for transmission. Person-to-person spread is usually the result of faecal-oral transmission of the infective stage. Indirect spread is also possible—eg, food or water may be a vehicle if hygiene is poor. There have been few reports of foodborne and waterborne protozoal disease during the past 40 years. In a review of foodborne disease in the USA[1] (selected years 1952–82), only one incident of protozoal disease was recorded—namely, 5 cases of *Entamoeba histolytica* infection in 1967; no further details were given. From 1983 to 1987, four outbreaks of foodborne giardiasis were reported in the USA involving some 51 cases.[2,3] In 1991, a fifth outbreak in Turkey was recorded.[4] One incident of amoebiasis was reported in a review of waterborne disease in the UK (1937–86)—namely, an outbreak with 28 cases and 26 symptom-free excretors in 1950.[3] According to Galbraith and colleagues,[5,6] foodborne or waterborne giardiasis in England and Wales is also rare with only one documented incident. Foodborne protozoal infection is almost certainly under-detected, possibly by a factor of ten or more, and an aetiological agent is identified in fewer than half of all outbreaks of suspected foodborne infection.[1,2] Like viruses, the protozoa I describe here are obligate parasites and cannot multiply in food; nonetheless, foodborne transmission of viruses is recognised and outbreaks are now more commonly reported. Transmission of parasitic infections by means of potable water supplies in developed countries is increasingly being recognised.[7]

Table 15.1   Protozoa parasitic for man transmissible by water or food

Parasite	Pathogenicity for man	Stage transmitted
Balantidium coli	+	Cyst
Blastocystis hominis	−/+	Not known
Cryptosporidium parvum	+	Oocyst
Other *Cryptosporidium* spp	−/+★	Oocyst
Chilomastix mesnili	−/+	Cyst
Dientamoeba fragilis	−/+	Trophozoite
Endolimax nana	−/+	Cyst
Entamoeba coli	−	Cyst
Entamoeba histolytica	+	Cyst
Microsporidia (eg, *Enterocytozoon*)	−/+	Spore
Enteromonas hominis	−	Cyst
Giardia intestinalis	+	Cyst
Iodamoeba butschlii	−	Cyst
Isospora belli	−/+	Oocyst
Retortamonas intestinalis	−	Cyst
Sarcocystis spp	+	Oocyst/tissue stages
Toxoplasma gondii	+	Oocyst/tissue stages
Trichomonas hominis	−	Trophozoite
Trichomonas tenax	−	Trophozoite

−/+ = low or doubtful for immunocompetent individuals; pathogenicity for the immunocompromised individual may differ.
Compiled from data in refs 8–11, 14, 18.
★Transmission to man probably occurs but seems to be uncommon.

Natural history

When protozoa with a simple life cycle are ingested they develop fully in a single host (monoxenous). Alternatively, in some species with a complex life cycle, ingestion of an oocyst or of an endogenous (tissue) stage in infected meat is an essential biological stage in the development cycle which sometimes involves more than one host species. In the definitive or final host, there is a sexual stage, and in the intermediate host, an asexual stage of parasite development. Examples of these developmental patterns can be found among protozoal infections that are potentially transmissible to man by food and potable water. There are at least three monoxenous genera of enteric protozoa with species pathogenic for man—namely, *Giardia*, *Entamoeba*, and *Cryptosporidium*—which can be transmitted by the faecal-oral route via food or water.[8-10] Other species may do so but are either rare or are believed to have low or no pathogenicity for man[9-11] (Table 15.1). Protozoa for which ingestion of infected meat by an alternative host is part of a developmental stage or, in some cases, a biological "dead end", include *Toxoplasma* and *Sarcocystis*.[12-14] All recognised waterborne protozoal enteric infection in man originates from environmental contamination with human or animal faeces. Vegetables and fruit may likewise be contaminated by use of polluted water for irrigation or during culinary preparation. Prepared foods may become contaminated by foodhandlers who are excreting infective stages. However, direct detection of parasites in food is difficult or impossible since there is no equivalent of the bacterial enrichment culture system that is suitable for recovery of small numbers of organisms from food; evidence that protozoa have a causal role is usually by epidemiological association: microbiological confirmation is usually lacking.

Foodborne enteric infection

GIARDIASIS

Giardia spp are binucleate flagellate protozoa with trophic (feeding) and cystic (resting) stages found in various mammals, birds, and other species—eg, amphibians and reptiles. There are believed to be at least 40 species of *Giardia*, but definitive characterisation of many of them is lacking. *G intestinalis* (formerly *G lamblia*), the species known to infect man, was first discovered by Leeuwenhoek (1681) in his own stools but was first properly described by Lambl (1859) (Fig 15.1).

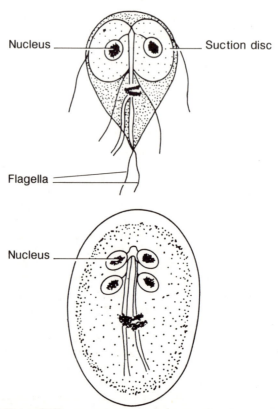

Nucleus — — Suction disc

Flagella

Nucleus

Figure 15.1 *Giardia intestinalis*
Trophozoite (top) and cyst (bottom)

Infection follows ingestion of viable cysts; the incubation period is usually 1–3 weeks and infection lasts, if untreated, usually for 4–6 weeks. Symptomless carriage is common, symptoms occurring in as few as 20% of infections. The infection is principally of the upper small bowel; the parasite preferentially inhabits the mucosal surface but sometimes becomes superficially invasive. Symptoms include diarrhoea with pale offensive stools, steatorrhoea, abdominal cramps, bloating, and sometimes pronounced fatigue and weight-loss; there may be malabsorption, especially of fats and fat-soluble vitamins.[14] Diagnosis is usually by detection of cysts and/or trophozoites in stools by direct microscopy (with or without concentration), or by immunofluorescence or enzyme-linked immunosorbent assay. Duodenal and jejunal aspiration or the "string test" may reveal parasites when stool examination is negative. However, demonstration of the parasite is not conclusive proof of causal association with symptoms. Acute gastrointestinal infection with other agents, or even non-infective enteropathy, may lead to the chance recognition of previously undetected carriage of the parasite. Trophozoites may be excreted during the acute diarrhoeal stage but they are unlikely to be infective because they do not remain viable for long and would not usually survive the effects of gastric acid. The cysts may be excreted intermittently in the stools of untreated cases for weeks or months, although the numbers that are excreted fluctuate. Cysts can be found in sewage effluents, in surface waters, and in some potable water supplies, occasionally even in developed countries.[15] Cysts can withstand chlorination at concentrations used to disinfect water; they may survive for longer than 2 weeks in a cool moist environment but are susceptible to heat and probably to prolonged freezing, though ice used in drinks was implicated as the source of infection in some cases. The infectious dose is believed to be low (≤ 10 cysts).[8–10,14]

Infection with *G intestinalis* has been increasingly recognised in the past few years and is now the most commonly reported protozoan infection world wide.[14] It is generally associated with poor hygienic conditions, including poor control of water quality.[8,10,14] Infection is common among children who attend day-care centres but is also associated with travel, particularly to Leningrad in the USSR.[6,10,14–17] Cysts have been detected in salad, fruit, and other food items; this finding and the association of infection with infected foodhandlers or with potable water indicate their potential role in transmission.[10,18] Outbreaks have been associated with potable water supplies in both the UK and the USA.[15,18,19] There are four known food-associated outbreaks. In the first,[20] at a school in Minnesota, USA, 29 cases were associated with consumption of salmon and a cream-cheese dip. The wife of a member of staff had prepared some of the food; she had changed the nappy (diaper) of her 12-month-old grandson just before food preparation. Although symptom free, the child was excreting cysts of *Giardia* in large numbers. In the second episode,[21] giardiasis developed in 13 of 16 people who ate at a picnic. The vehicle of infection seemed to be a cold noodle salad. A further case occurred in an individual who did not attend the picnic but who subsequently ate the salad at home. Spices were added by hand to the noodles by a neighbour who became ill the next day. Her children who had not attended the picnic, were subsequently shown to be excreting cysts of *Giardia*. In the third outbreak, in a nursing home in Minnesota, USA, there were multiple modes of transmission, including food, for 88 cases;[22] there was an association with food for staff who ate sandwiches but not for patients who ate cooked food only. The fourth outbreak, in New Jersey, USA, was associated with the consumption of a fruit salad at a family party; 10 out of 25 people became ill.[3] The woman who prepared the fruit salad had a child in nappies and a pet rabbit, both of whom were excreting *Giardia*. In the most recent outbreak in Turkey,[4] which affected two families, soup prepared from sheep tripe was thought to be implicated.

Campers (back-packers) in the USA are thought to have acquired zoonotic *Giardia* infection from consumption of raw pristine waters contaminated by wildlife, particularly beavers.[14,22] An outbreak in Bristol, England, with 108 confirmed cases, was believed to be due to post-treatment contamination of the potable water supply.[18]

AMOEBIASIS

Entamoeba are anaerobic amoebae that are mainly commensals of the gut lumen of various host species. Infection follows ingestion of viable cysts with a clinical incubation of a few days to many months (commonly 2–6 weeks). Cysts develop from trophozoites within the gut lumen before being passed in the faeces. Nuclear division takes place within the cyst. Further development can occur in the cyst excreted into the environment. Nuclear sub-division leads to the release, on excystment, of numerous trophozoites. *E histolytica*, the human intestinal amoeba, was first described by Losch (1875) in St Petersburg (Leningrad). The so-called "small-race" of *E histolytica* is a harmless commensal now known as *E hartmanni*.[14,23]

Infection with *E histolytica* is usually associated with tropical dysentery with bloody or mucoid diarrhoea, and in some cases with systemic spread and disease. Most infections, however, are symptomless or produce only mild bowel disturbance. Diagnosis is by detection of trophozoites or cysts in fresh stools. Serological diagnosis may be useful in systemic disease. Strains or isolates of *E histolytica* differ with respect to their pathogenicity and this correlates with their isoenzyme (zymodeme) profiles.[14,23,24]

Infection is generally associated with poor hygiene and poor water quality and is endemic in many poor communities in both temperate and tropical regions. Infection is transmitted by the environmentally resistant cyst stage and is derived, directly or indirectly, from other human cases: there is no known animal reservoir, although other host species can be infected.[23] Transmission by food and water was described in the 1940s, especially as a result of excretion by foodhandlers.[8-10] Several waterborne outbreaks have been recorded both in the USA and in the UK. A UK outbreak at an airforce base in 1950 was due to sewage contamination of a borehole water supply.[5] Infection was probably introduced by servicemen who had been infected in the tropics, although indigenous infection has been reported in the UK.[23] The organism can be transmitted to food by flies.[10,23] The infection is also part of the travellers' diarrhoea syndrome, in which food and water are probably important, and of the "gay bowel syndrome", transmitted by homosexual practices, in which strains of low pathogenicity seem to be commonly involved.[8,14,23]

There has been a pronounced decline in reports of amoebiasis in the UK in the past 20 years (Public Health Laboratory Service, Communicable Disease Surveillance Centre, unpublished).

CRYPTOSPORIDIOSIS

Cryptosporidium spp are coccidian protozoa with at least 4 confirmed species—2 in mammals (*C muris* and *C parvum*) and 2 in birds (*C meleagridis* and *C baileyi*). There are more than 40 host species from which *Cryptosporidium* spp have been isolated, although some isolates are poorly described. Some so-called "species" of *Cryptosporidium* are probably *Isospora* or *Sarcocystis* spp, sporocysts of which have been mistaken for oocysts of *Cryptosporidium*. Other "species" are synonyms for *C parvum*. *Cryptosporidium* has a worldwide distribution.[6,14–16,19,25–27] Most human infection is probably due to *C parvum*; infection with this species is also common in livestock animals, especially cattle and sheep. Pigs, goats, and horses can also be infected. There is, therefore, a large zoonotic potential reservoir of infection. Pets are occasionally infected but they do not seem to be an important source of the organism. The avian cryptosporidia seem to be transmissible poorly, if at all, to man and other mammals.[19,25,27]

The parasite is monoxenous. After ingestion of the transmissible stage (the

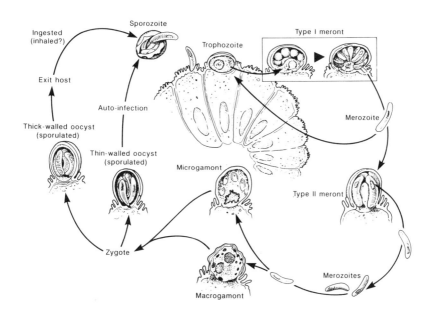

Figure 15.2 Life cycle of *Cryptosporidium parvum*.
From Current WL, Blayburn BL. Cryptosporidium infections in man and domesticated animals. In: Long PL, ed. Coccidiosis of man and domestic animals. Boca Raton: CRC Press, 1990: 159. Adapted from the original drawing by Kip Carter, University of Georgia.

oocyst), the infective sporozoites are released in the gut where they undergo asexual development in a pseudo-external intracellular but extracytoplasmic location, mainly in the apical enterocytes of the small bowel. After two such stages, differentiation takes place into sexual stages with the production of zygotes and more oocysts. The oocysts mostly develop (sporulate) within the gut and are excreted in a fully infective form that does not need the further maturation (ripening) period usually found in coccidia (Fig 15.2).[19,25] The incubation period after ingestion is about 3–11 days, although wider limits have been suggested.[19] The symptoms include a secretory-type watery diarrhoea, vomiting, anorexia, malaise, and weight-loss. There is currently no fully effective specific treatment.

Cryptosporidium is a common cause, world wide, of infection in man—ie, an acute and often persistent gastroenteritis both in immunocompromised individuals, especially those with AIDS, and in otherwise healthy people.[19,25–27,29] The natural history of cryptosporidiosis is complex and involves both zoonotic and "urban" (non-zoonotic) reservoirs and modes of transmission. Assumed initially to be a zoonosis, cryptosporidiosis can often be transmitted directly from person to person and indirectly by fomites, water, and sometimes food.[11,19,25–29] As with other enteric protozoa, there is a potential for transmission by foodhandlers who are excreting oocysts. The observation that excretion by most patients seems to stop within a moderately short period of clinical resolution,[19] is based on insensitive microscopic detection methods. Family outbreaks are not uncommon,[19,26] in some of which food is likely to have been the vehicle of transmission. In the early 1980s attention was drawn to sporadic infection in the community, especially in otherwise healthy children. The pattern of infection suggests that the illness is endemic in many of the areas from which these reports came.[11,19,25–27] The infection is also an important cause of travellers' diarrhoea. That such cases often have mixed infections, especially with *Giardia*, points to a common epidemiology which may involve contaminated water or food. Increased exposure to livestock may be relevant in some areas. Direct and indirect person-to-person transmission is also important in view of the generally poor hygienic conditions in many lesser-developed countries. Infection in children in some Scandinavian countries seems to be uncommon; the parasite is found mainly in adults, most of whom acquired their infection while travelling abroad, especially to the USSR, or by occupational exposure to livestock.[16,19,25–27]

Several studies have noted temporal or seasonal peaks in incidence of cryptosporidiosis, which in some cases coincided with periods of maximum rainfall and with lambing and calving.[19] The widespread practice, even in advanced countries, of disposal of both animal and human excreta to land—eg, by muck and slurry spreading on pasture—may lead to infection by contamination of water courses and reservoir feeder streams. Pollution of surface waters naturally or by these practices may lead to contamination of water supplies or of food crops during irrigation. Oocysts have been found in water (including surface waters and in a few instances potable supplies) in the UK and in North America and there have been cases and outbreaks in which water, including potable supplies, was the means of transmission.[7,15,19,28]

Direct incrimination of food in the transmission of *Cryptosporidium*, as with other protozoa, is hampered by the lack of the bacteriological equivalent of enrichment culture for recovery of small numbers of oocysts and for confirmation of viability. In-vitro cultivation in tissue culture and in fertile hens' eggs is possible but does not result in amplification.[25] Epidemiological evidence points to the consumption of certain foods (especially raw sausages, raw offal, and raw milk) as a risk factor.[6,19,26] Outbreaks due to contamination of food by foodhandlers excreting oocysts have not yet been recorded.

OTHER INFECTIONS

Balantidium coli. Balantidia are found in various host species, especially as a commensal in pigs, and are distributed world wide. These large (70 μm diameter) actively motile ciliated protozoa multiply by binary fission and preferentially inhabit the large bowel.[14] The environmentally resistant protective cysts which are excreted in the stools are not an essential developmental stage. Infection in man follows ingestion of cysts which can result in symptomless luminal colonisation. Invasive infection of the colonic mucosa can lead to colitis. There may be diarrhoea (which is sometimes dysentery-like), abdominal colic, tenesmus, nausea, and vomiting. The incubation period is unknown but is probably a few days. Diagnosis is by detection of the large ciliated trophozoites or cysts in the stools. Infection can be transmitted from person to person or via food and the cysts of the parasite can survive in water. Human infection with this protozoan seems to be rare.[9–11] Pigs may be infected with either *B suis*, which is not infectious for man, or *B coli*. Although porcine strains of *B coli* are said to be poorly adapted to man, human infection is usually associated with close contact with pigs, including shared habitation.[14] Infection is uncommon in Muslims, even in poor hygienic conditions, a finding also seen with *Cryptosporidium* in some studies.[19]

Isospora belli. The isospora are monoxenous coccidian parasites that inhabit the small bowel. Little is known about the endogenous stages in man but oocysts are excreted largely unsporulated and need a period of ripening when they divide internally to produce two sporocysts, each containing four sporozoites. Human infections have been reported mainly from tropical and sub-tropical regions but the prevalence generally is unknown. The incubation period has not been defined but is presumably similar to that for cryptosporidiosis. Diagnosis is by detection of oocysts in the stools. *Isospora* infection is particularly associated with AIDS patients with diarrhoea.[29] In immunocompetent individuals most infections are probably harmless or lead to mild transient diarrhoea. The origin of most infections is unknown but *Isospora* spp are believed to be host-specific and transmission is probably from person to person directly[14] or via food or water.[9,29] There is confusion about classification, which may affect diagnosis, because cryptosporidial oocysts are similar to the sporocysts of *Isopora* and *Sarcocystis* spp.[12,14]

Sarcocystis spp. These organisms are coccidian obligate two-host parasites: there is a carnivore definitive host (eg, cats, dogs, or man), in which sexual development takes place with production of oocysts, and an intermediate host (eg, cattle, sheep, pigs), in which asexual cysts are formed in the tissues.[12,14] Of the dozen or more species, 2 infect man—*S hominis*, which infects cattle, and *S suihominis* from pigs. Although common in livestock, infection in man is rare, which may reflect low transmissibility to man, or susceptibility of the organism to cooking and freezing, or may suggest that infection is commonly subclinical. The incubation period is unknown. Symptoms, which are generally mild with nausea, diarrhoea, and malaise, develop within a few hours of ingestion. Diagnosis depends on detection of oocysts in stools.

Chilomastix mesnili. This is a cosmopolitan flagellate parasite found in the large bowel in man, other primates, and pigs. Transmission is probably via contaminated food or water and mild symptoms of diarrhoea may result from infection. The incubation period is unknown.

Blastocystis hominis. An organism of uncertain taxonomic status, *B hominis* is generally regarded as a protozoan, although it has also been referred to as a yeast:[14,30] it may be able to cause diarrhoea, although this is doubtful in healthy individuals.[30] The role and possible importance of this organism has lately been discussed.[31] Transmission is probably by the faecal-oral route and thus food is a potential vehicle of transmission.

Microsporidia (Microsporans). Microsporidia, including *Enterocytozoon* spp, are obligate intracellular protozoa found in many host species and have been recognised in some AIDS patients. Their importance and mode of transmission is currently uncertain.[14,32] Detection and identification are very difficult. Transmission is probably by means of a spore stage in the stools of infected subjects as described by van Gool and colleagues.[33]

Foodborne systemic infection

TOXOPLASMOSIS

The coccidian genus, *Toxoplasma*, contains a single species, *T gondii*; it is an obligate intracellular parasite to which up to nearly half the adult population in the UK[34] and 30–40% of adults in the USA[12,13] have demonstrable antibody. For many years there was doubt about the existence of an oocyst stage. However, in the definitive host (members of the cat family), a sexual stage of the cycle in the intestines leads to the production of oocysts, which are shed in the faeces in large numbers. The oocysts, which require a period of extrinsic maturation (sporulation), can survive under appropriate conditions in the environment for months.

Infection in man usually follows ingestion of viable oocysts with an incubation period believed to be 5–20 days. Person-to-person transmission is known only from mother to child in utero. Infection is probably acquired by cats from wild mice, birds, or from being fed raw meat or by the faecal-oral route from other cats.[9,12,13,34] In other host species, ingestion of oocysts results in initiation of an asexual cycle with the formation of tissue cysts (bradycysts). Ingestion of these tissue cysts by any vertebrate leads to release of the motile bodies (bradyzoites) from within the cyst by the action of digestive enzymes; the motile bodies then initiate fresh asexual cycles in an acute infection. Infection in the intermediate host can be acquired by vertical transmission. Infection in children and adults is symptomless or is characterised by a mild influenza-like illness or lymphadenopathy: congenital infection may lead to severe defects. Infection is serious in immunocompromised people, such as transplant recipients and AIDS patients, in whom reactivation of latent infection may occur.[9,13,34]

Herbivores, including livestock animals, can become infected by ingestion of grass and other feedstuffs that have been contaminated by cat faeces. Their tissue may remain infectious for life.[13]

Foodborne infection in man is probably uncommon.[34] The tissue cysts are readily killed by heat and, albeit more unreliably, by freezing.[12] Raw meat dishes are a possible source of infection—eg, steak tartare—although beef is uncommonly infected. When infection has been demonstrated, other meats (such as pork, mutton, or horse meat) have been implicated. A family outbreak of toxoplasmosis in Australia was thought to be associated with consumption of home-made Lebanese kibi—a meat dish traditionally made with raw lamb.[35]

SARCOCYSTOSIS

Sarcocysts have been occasionally detected in the muscle (striated and cardiac) of man in biopsy and necropsy samples, which is an indication of systemic spread. It has been suggested that this may be due to infection with a species for which man is an intermediate host, the definitive host being an unidentified carnivore.

AMOEBIASIS

Systemic haematogenous dissemination of *Entamoeba histolytica* may result in abscess of the liver or, less commonly, of the lung, brain, or other organs, and in ulceration of the perineum. Amoebic granulomas in the wall of the gut may be mistaken for carcinoma.[23]

Control of infection

Protozoal cysts and other life-cycle stages are moderately sensitive to heat, to prolonged freezing, and, in some cases, to desiccation. Cysts/oocysts are moderately resistant to disinfection (eg, chlorination); in a cool moist environment, they may survive for weeks or months. Recent studies have shown that cryptosporidial oocysts are sensitive to ozone in concentrations that may be practicable for use in water treatment, and to chlorine dioxide.[7]

Control depends primarily on breaking the chain of infection—eg, by preventing contamination of food and water or of food by the hands of excretors during preparation: good personal hygiene must be observed while symptoms persist.[36] There are no reliable criteria of clearance for enteric protozoa such as *Giardia*. Where sewage disposal is unsatisfactory, flies may be important in transmission and their control is essential.[20,23] Chronic, sub-clinical, or symptomless carriage of an organism may be discovered when faeces are examined as a result of acute super-infection with other pathogenic enteric agents. The presence in stools of some of the parasites of little or no pathogenicity, such as *Enteromonas* and *Entamoeba coli*, particularly when mixed infections are present, may indicate poor hygienic conditions and the potential for foodborne or waterborne transmission of pathogenic species. Some species of enteric amoebae and trichomonads lack a cystic stage and are generally thought to be transmitted directly from person to person. Transmission of such trophic parasite forms can, however, be enhanced by ingestion in food, which provides protection during passage through the stomach and small intestine. They are of little clinical consequence.

When more than one host is involved in the life cycle, and if veterinary control is not possible, reliance must be placed on adequate cooking of potentially infected meat. In toxoplasmosis, infection both with faecal oocysts and with tissue cysts is possible and care should be taken, especially by pregnant women, not to handle contaminated cat litter before food preparation and to take care in the handling of raw meat.

The development of more sensitive methods that use immunological and molecular-based systems may help to provide further information about protozoal foodborne infections, their detection, and control.[8,11–13,19,23,24]

I thank Dr D. Warhurst, Dr Joan Rose, and Dr R. Fayer for help and advice.

References

1. Cliver DO. Foodborne disease in the United States, 1946–1986. *Int J Food Microbiol* 1987; **4**: 269–77.
2. Bean NH, Griffin PM, Goulding JS, Ivey CB. Foodborne disease outbreaks, 5-year summary, 1983–87. *MMWR* 1990; **39**: 15–57.
3. Porter JDH, Gaffney C, Heymann D, Parkin W. Food-borne outbreak of *Giardia lamblia*. *Am J Public Health* 1990; **80**: 1259–60.
4. Karabibers N, Aktaş F. Foodborne gardiasis. *Lancet* 1991; **337**: 376–77.
5. Galbraith NS, Barrett NJ, Stanwell-Smith R. Water and disease after Croyden: a review of water-borne and water-associated disease in the UK 1937–86. *J Instit Water Environ Manag* 1987; **1**: 7–21.
6. Galbraith NS, Barrett NJ, Sockett PN. The changing pattern of foodborne disease in England and Wales. *Public Hlth* 1987; **101**: 319–28.
7. Cryptosporidium in water supplies. Report of the group of experts. (Chairman: Sir John Badenoch.) London: HM Stationery Office, 1990.
8. Ravdin JI, Weikel CS, Guerrant RL. Protozoal enteropathies: cryptosporidiosis, giardiasis, and amoebiasis. In: Guerrant RL, ed. Clinical tropical medicine and communicable diseases. Diarrhoeal Diseases. London: Bailliere Tindall, 1988: no 3, 503–36.
9. Fayer R, Gamble HR, Lichtenfels JR, Bier JW. Foodborne and drinkborne parasites. In: Vanderzant C, Splittstoesser DF, eds: American Public Health Association's compendium of methods for the microbiological examination of foods. Washington, DC: American Public Health Association (in press).
10. Barnard RJ, Jackson GJ. *Giardia lamblia* The transfer of human infections by foods. In: Erlandsen SL, Meyer EA, eds. *Giardia* and giardiasis. New York: Plenum Press, 1984: 365–78.
11. Mata L. *Cryptosporidium* and other protozoa in diarrheal disease in less developed countries. *J Pediatr Infect Dis* 1986; **5** (suppl): S117-S130.
12. Fayer R, Dubey JP. Methods for controlling transmission of protozoan parasites from meat to man. *Food Technol* 1985; **39**: 57–60.

13. Murrell KD, Dubey JP, Fayer R. Role of biotechnology in the control of foodborne parasites. In: Bills DD, Kung S-D, eds. Biotechnology and food safety. Proceedings of the second International Symposium, 1989. University of Maryland. Massachusetts: Butterworth, 1990: 215–25.
14. Warhurst DC, Green EL. Protozoal causes of diarrhoea. *PHLS Microbiol Dig* 1988; 5: 31–37.
15. Rose JB, Gerba CP, Jakubowski W. Survey of potable water supplies for *Cryptosporidium* and *Giardia*. *Environ Sci Tech* (in press).
16. Jokipii AMM, Hemilä M, Jokipii L. Prospective study of acquisition of *Cryptosporidum, Giardia lamblia,* and gastrointestinal illness. *Lancet* 1985; ii: 487–89.
17. Brodsky RE, Spencer HC, Schultz MG. Giardiasis in American travelers to the Soviet Union. *J Infect Dis* 1974; 130: 319–23.
18. Jephcott AE, Begg NT, Baker IA. Outbreak of giardiasis associated with mains water in the United Kingdom. *Lancet* 1986; i: 730–32.
19. Casemore DP. Epidemiological aspects of human cryptosporidiosis. *Epidemiol Infect* 1990; 104: 1–28.
20. Osterholme MT, Forfang JC, Ristinen TL, et al. An outbreak of foodborne giardiasis. *N Engl J Med* 1981; 304: 24–28.
21. Petersen LR, Cartter ML, Hadler JL. A foodborne outbreak of *Giardia lamblia*. *J Infect Dis* 1988; 157: 846–48.
22. White KE, Hedberg CW, Edmonson LM, Jones DBW, Osterholm MT, MacDonald KL. An outbreak of giardiasis in a nursing home with evidence for multiple modes of transmission. *J Infect Dis* 1989; 160: 298–304.
23. Warhurst DC. *Entamoeba histolytica* and amebiasis. In: Feachem RG, Bradley DJ, Garelick H, Mara DD, eds. Sanitation and disease. Health aspects of excreta and wastewater management. Chichester: John Wiley, 1983: 337–47.
24. Chiodini P. Parasitology. In: Reeves DS, Geddes AM, eds. Recent advances in infection: 3. Edinburgh: Churchill Livingstone, 1989: 237–46.
25. Current WL. *Cryptosporidium*: its biology and potential for environmental transmission. *CRC Crit Rev Environ Control* 1986; 17: 21–51.
26. Casemore DP, Palmer SR, Biffin A. Cryptosporidiosis. *PHLS Microbiol Digest* 1987; 4: 1–7.
27. Fayer R, Ungar BLP. *Cryptosporidium* spp and cryptosporidiosis. *Microbiol Rev* 1986; 50: 458–83.
28. Smith HV, Rose JB. Waterborne cryptosporidosis. *Parasitol Today* 1990; 6: 8–12.
29. Soave R, Johnson WD. *Cryptosporidium* and *Isospora belli* infections. *J Infect Dis* 1988; 157: 225–29.
30. Miller RA, Minshew BH. *Blastocystis hominis*: an organism in search of a disease. *Rev Infect Dis* 1988; 10: 930–38.
31. Editorial. *Blastocystis hominis:* commensal or pathogen? *Lancet* 1991; 337: 521–22.
32. Shadduck JA. Human microsporidiosis and AIDS. *Rev Infect Dis* 1989; 11: 203–07.
33. van Gool T, Hollister WS, Schattenkerk JE, et al. Diagnosis of *Enterocytozoon bieneusi* microsporidiosis in AIDS patients by recovery of spores from faeces. *Lancet* 1990; 336: 697–98.
34. Fleck DG. Toxoplasmosis. *PHLS Microbiol Dig* 1989; 6: 69–73.
35. De Silva LM, Mulcahy DL, Kamath KR. A family outbreak of toxoplasmosis: a serendipitous findings. *J Infect* 1984; 8: 163–67.
36. Anon. Communicable Disease, Supplement: Notes on the control of human sources of gastrointestinal infections, infestations and bacterial intoxications in the United Kingdom. Communicable Disease Report, Supplement no 1. London: Public Health Laboratory Service, Communicable Diseases Surveillance Centre, 1990.

16
Natural foodborne toxicants

M. R. A. Morgan and G. R. Fenwick

In 1971, Wodicka[1] used onset, severity, and prevalence of symptoms to rank food hazards in order of importance. He regarded microbiological contamination as the most important risk, closely followed by nutritional imbalance; the third was environmental contamination, followed by natural toxicants. Of least importance according to these criteria are risks associated with agrochemicals and food additives. Hall[2] has suggested that there is a factor of 10^5 between these extremes of risk. Even though there is no scientific evidence to cast doubt on this grading it would be appropriate now to repeat this ranking exercise with information from studies that have been done during the past 20 years. An additional requirement of such a study would be to "anchor" the ranking against the risks of alternative hazards—eg, those associated with death from heart disease or traffic accident, to which the general public is, arguably, more able to relate.

Food is the most chemically complex substance commonly encountered by the public. There are probably more than half a million naturally occurring compounds in fresh plant foods; and more are formed as a result of domestic food preparation or industrial processing. These natural compounds are responsible for the texture, appearance, and flavour of plant foods, as well as for their nutritional value and physiological effects. By contrast, there are less than a few thousand recognised and approved food additives and agrochemicals; and these have been subjected to rigorous biological evaluation.[3] Such toxicological scrutiny is almost unknown for naturally occurring food compounds, even for those that clearly possess biological activity. Available data commonly relate to the results of in-vitro procedures or animal bioassays, the relevance of which to man is unclear. Furthermore, studies are hampered by the difficulty of obtaining the compound of interest in sufficient amount and purity for detailed biological examination. In view of these facts, it is perhaps not surprising that Wodicka placed the risks due to natural toxicants well ahead of those associated with synthetic chemicals entering the food chain.

Thus, there is a substantial number of naturally occurring dietary compounds about which there is either incomplete evidence or reasonable suspicion of the occurrence of long-term effects in man. It is noteworthy that if these compounds were to be presented as synthetic chemicals to the appropriate government committees, their use in food would not even be considered.

Public perception and risk/benefit

Many surveys have confirmed that the public associates "natural" with healthy, wholesome, and safe. The active ingredients of natural products and herbal remedies are perceived to be unrelated to chemicals, which in turn are commonly regarded as man-made, damaging to the environment, and, in the context of their presence in food, hazardous and frequently carcinogenic. However, the simple facts are that (i) food consists of chemicals (both natural and synthetic); (ii) much more is known about the biological effects of synthetic chemicals; (iii) natural is not necessarily good for you; and (iv) public concern and scientific investigation should be based on the bioactivity rather than on the origin of the chemical.

To make a risk/benefit assessment for an individual chemical, information is needed about the concentrations in the diet so that calculations on exposure can be made. In view of the diversity of diets world wide it is important that such calculations are based both on the general population and on certain groups that are judged to be especially at risk. This information is not yet available for all compounds of interest. Table 16.1 shows information about the exposures of the UK population and individual sub-groups to various biologically active plant constituents. We have included these data merely to show the possible population-variation of such exposure and to highlight the need for complementary clinical/toxicological data.

Although there is plenty of information about the biological effects of chemicals isolated from plants, the quality of such data and relevance to man are often dubious Generally, there is a lack of biological (toxicological and clinical) information about potency of dietary compounds in man. Such information, with chemical data on intake, is necessary for proper risk/benefit assessment and without such assessment little can yet be said about the importance of such compounds for human health.

Table 16.1 UK mean daily intake of natural toxicants

Class of compound	Population	
	Total	Vegetarian
Glucosinolates (brassicas)	50	110
Glycoalkaloids (potatoes)	13	70–90
Saponins (legumes)	15	100 (*220)
Isoflavones (soya)	<1	105

*UK vegetarian population of East African origin.
Values in mg.

Plant toxicants

We have chosen a limited selection of examples of plant toxicants to illustrate the range and nature of symptoms. For more detailed information, we refer the interested reader to some general sources.[4–9]

Cassava is a staple dietary item in tropical Africa, South America, and Southeast Asia. The plant is a rich source of cyanogenic glycosides, which may be hydrolysed chemically or enzymically to yield hydrogen cyanide. Careful processing—namely, soaking, fermentation, and drying—is therefore vital to reduce to a minimum serious illness or to prevent death. Fresh whole cassava root may contain the equivalent of 90–400 mg hydrogen cyanide/kg; the peel is the richest source. Signs and symptoms of acute cassava poisoning are dyspnoea, gasping, paralysis, coma, and death. Whereas concentrations of cyanide below 100 mg/kg in cassava products have not been associated with acute poisoning, tropical ataxic neuropathy (characterised by lesions of the optic, auditory, and peripheral nerves) is associated with dietary exposure to cassava products containing as little as 20–60 mg cyanide/kg. The major pathway for cyanide detoxification involves rhodanase catalysis, yielding thiocyanate, which inhibits intrathyroidal transfer of iodine and increases thyroid stimulating hormone levels. Increased incidence of goitre and cretinism may thus be associated with chronic dietary exposure to cassava and other cyanide producing plant species, particularly when iodine consumption is low. Cyanogenic glycosides in legume pulses may be especially troublesome because, like cassava, they can be a dietary staple. Restrictions have been applied in the USA against the introduction of certain varieties of lima bean which contain very high concentrations of cyanide precursors. In other countries, there have been cases of severe illness, and even death, due to consumption of cyanide-rich apple seeds, bitter almonds, and laetrile—the bogus cancer remedy produced from peach kernels. Thiocyanate may also be formed via breakdown of another family of plant glycosides, the glucosinolates, which are found in cruciferous vegetables. Cabbage and Brussels sprouts also contain a more potent goitrogen, 5-vinyloxazolidine-2-thione, the effects of which on the thyroid gland are not reversed by dietary iodine supplementation. Although brassica-induced goitre has been described in central and southern Europe and north Africa, elsewhere it is believed that the contribution of dietary goitrogens to thyroid dysfunction and disease is negligible.

An example of the role of epidemiological investigation to assess the association between diet and disease is provided by Jamaican vomiting sickness. Anecdotal evidence often cited an association between consumption of unripe ackee fruit and the subsequent onset of symptoms, including vomiting, drowsiness, and muscular weakness, followed by, in extreme cases, coma and death. According to chemical and biological studies the illness is due to a metabolite of the plant constituents, hypoglycin and γ-glutamylhypoglycin, the concentrations of which are greatest in the unripe fruit, and decrease greatly upon ripening. Currently, the US Food and Drug Administration prevents entry of ackee fruit and produce into the USA unless accompanied by information showing that the product is safe. Ackee poisoning is characterised by the onset of symptoms 4–10 h after eating the fruit. Onset of symptoms after consumption of bitter squash zucchini and other cucurbits may be even more rapid. Although cucurbits (including cucumber) are widely cultivated and are generally harmless, chance pollination or mutation will occasionally throw up a bitter "sport", which is visually no different from the normal plant. These sports contain increased levels of cucurbitacins, which are among the most bitter of natural plant substances, with one, cucurbitacin E, being detectable at a dilution of $1/10^8$. In view of this characteristic it is perhaps surprising that any cases of poisoning occur at all. That they do so is partly explained by the acute toxicity of these compounds. In one outbreak in Australia, severe cramps, diarrhoea, and collapse followed within 1–2 h of eating as little as 3 g of bitter zucchini. In a more recent poisoning in the UK, the bitterness may have been disguised when the zucchini were included in a meat casserole.

Legumes also contain a wide range of anti-nutritional and biologically-active components; these include oestrogenic isoflavones and coumestans (linked with reproductive disturbance in mammals), saponins and haemagglutinins (which affect the mucosal lining of the lower intestine), phytate and polyphenolics (also known as tannins) (which reduce availability of minerals such as iron, zinc, and calcium), oligosaccharides (which cause excessive intestinal gas production), and enzyme inhibitors. Fava beans are a major protein source for the Middle and Far East and north Africa and are a hazard for individuals who have a deficiency in glucose-6-phosphate dehydrogenase (G6PD). Such individuals cannot produce enough reduced glutathione to reduce endogenous oxidants generated by the hydrolysis products of the plant glycosides, vicine and convicine. Initial symptoms of favism include headache, nausea, vomiting, lumbar pain, and fever. Haemoglobinuria appears within 5–30 h, followed by jaundice; death usually occurs within 48 h but non-fatal attacks last 5–7 days. Favism accounts for 1–2% of paediatric hospital admissions in Cairo/Alexandria; the incidence in Sardinia is 5/1000. Although more than 100 million individuals world wide have a deficiency in G6PD, sensitivity to fava beans is associated with the Mediterranean variant. A widely distributed tropical legume, *Mucuna pruriens*, possesses seeds containing hallucinogenic N,N-dimethyltryptamine and other alkaloids. In Mozambique the seeds, consumed after boiling, are a common famine food. 203 cases of acute toxic psychosis were recently reported in a remote part of that country and were associated with this legume.[10] Shortage of water may have been a reason for incomplete detoxification of the seeds. That there was a preponderance of childbearing women (62%) and young girls (23%) in the outbreak may be due to their lowered micronutrient status and their habit of reserving the better food for the menfolk.

The potato glycoalkaloids are, perhaps, the most familiar natural toxicants because of the scale of potato production (currently over 250 million tonnes annually) and the common knowledge of the toxicity of green potatoes. Exposure of potatoes to light (which causes this greening due to chorophyll synthesis, unconnected to glycoalkaloid accumulation) is one of many stresses, such as fungal, mechanical, or insect damage, which together with sprouting lead to the accumulation of high concentrations of glycoalkaloids in the tuber. Although all parts of the potato plant contain these compounds, the parts above ground are richest (this is the reason for the toxicity of potato berries or apples). In the tuber, the peel contains the highest concentrations, so that normal food preparation will greatly reduce human exposure. Acute potato poisoning (eg, in Lewisham, London, 20 years ago[11]) is almost invariably associated with consumption of poor-quality damaged potatoes. In the Lewisham outbreak, 78 schoolchildren were poisoned; 17 required hospital treatment. Neurological symptoms incuded apathy, restlessness, drowsiness, and visual disturbance; the potatoes contained glycoalkaloid levels in excess of 330 mg/kg and had pronounced anticholinesterase activity. Commercial blemish-free tubers will generally contain less than 100 mg/kg, and this level is very substantially reduced after peeling. The recent finding that potato glycoalkaloids possess potent gut-permeabilising activity may point to the possibility of chronic effects of dietary glycoalkaloids;[12] any such effects will be enhanced in individuals who regularly eat large amounts of potato peel, for example, in the mistaken belief that this is a rich source of vitamins and fibre. The hypothesis originally proposed in the early 1970s, that potato glycoalkaloid consumption was associated with increased incidence of spina bifida and anencephaly, seems now to have little support.

Fish toxicants

Natural toxicants are not confined to plants; some of the most potent toxins are found in marine products. Periodically, large areas of the world's seas may become brightly coloured and luminescent due to the rapid growth of dinoflagellates and plankton, some of which contain highly toxic alkaloids. These "blooms" are frequently associated with increased mortality of fish and other animals, but pelecypods (ie, bivalve molluscs, such as clams, oysters, scallops, and mussels) can absorb the toxic components from plankton without harm. Subsequent consumption of such shellfish will produce immediate and severe effects, primarily on the central and peripheral nervous systems, known as paralytic shellfish poisoning. Initial symptoms are seen within 30 min of consumption—namely, tingling and numbness in the extremities, which spreads quickly throughout the body producing general lack of muscular coordination. Depending on the amount of toxin ingested (0·5–1 mg per adult may be lethal), death due to respiratory paralysis may occur within 12–24 h. An algal bloom was also responsible for an epidemic of toxic encephalopathy in Canada in 1987.[13] As a result, mussels harvested around Prince Edward Island contained domoic acid (30–120 mg per 100 g tissue), which is a heat-stable, neuroexcitatory aminoacid. Subsequent consumption of the toxic shellfish produced gastrointestinal symptoms and neurological dysfunction, chronic memory loss, and motor neuropathy or axonopathy. Following the outbreak, screening for domoic acid, with both chemical and biological methods, is routinely conducted before commercial distribution.

Ciguatera poisoning is endemic in many countries where large and regular amounts of fish from tropical waters are eaten. Although symptoms are often severe and diverse, including gastrointestinal and neurological disorders, skin rashes, psychological disturbances, and cardiovascular effects, mortality is generally low. Neurological disturbance may last from several days to many weeks, and any subsequent intoxication will lead to more severe symptoms. The origin of ciguatera poisoning was unknown for many years but analysis of the diets of toxic fish suggests that the toxins (heat-stable, fat-soluble compounds) originate in epiphytic microalgae, enter the food chain via the grazing of herbivores, and are distributed through a wide-ranging carnivore population. Thus, it has been claimed that over 400 species of fish, from almost 60 different families, are carriers of the ciguatera toxins, which occur at very low levels (typically <10 parts per billion). It has been suggested that the large increase in ciguatera poisoning in French Polynesia between 1960 and 1973 was partly due to the extensive reef damage (and resultant release of toxic microalgae) after the French nuclear testing programme in that region. Since it is impossible for ciguateratoxic fish to be identified by appearance and taste, dietary variation, avoidance of large reef predators, and keeping to local custom and advice seem to be among the most effective ways of reducing risk.

Unlike the wide diversity in origin of ciguatera poisoning, scombroid poisoning is primarily associated with consumption of mackerel, tuna, and related species, although anchovies have also been implicated. Typical symptoms include headache, palpitations, gastrointestinal disturbance, and generalised erythema and urticaria. Such symptoms may last 1–8 days; the fact that they are variable and that not everyone who consumes a toxic fish will be affected is due to the extreme variability of scombrotoxin within and between individual fish. Scombrotoxin is thought to be a complex mixture that is derived from the action of bacteria on histidine and related compounds present in fish skin and muscle. Thus, unlike the above examples, post-catch handling plays an important part in scombroid poisoning. Histamine production is greatly enhanced (up to 1000-fold) when fresh fish is stored at room temperature, rather than in ice. The role of histamine in scombroid poisoning has been questioned. Nonetheless, treatment with antihistamine agents, such as cimetidine, is effective. Although additional toxic substances may be present in fish (including perhaps histaminase inhibitors), efficient monitoring of histamine levels in fresh, canned, and processed fish is an effective quality control measure.

Mycotoxins

Although mycotoxins are usually contaminants of foods, especially as a result of poor storage, they merit inclusion with true natural toxicants because under many circumstances such contamination is almost inevitable. Mycotoxins are chemically very diverse; some are extremely potent and several are carcinogenic. Toxins are synthesised by particular fungi under certain environmental and growth conditions. Thus, *Aspergillus flavus* and *A parasiticus* can synthesise aflatoxins under high temperature and humidity; the production of trichothecenes by *Fusaria* spp or ochratoxin A by *A ochraceus* requires lower temperatures, such as those in northern Europe. Mycotoxins have been found in many types of foods for human consumption, particularly cereals and nuts. Contamination can occur in the field or during storage. Meat, eggs, and milk from animals that have been fed on mycotoxin-contaminated feed can provide an indirect route of exposure.

In view of the potency of some mycotoxins it is not surprising that incidents of mycotoxicoses can occasionally assume dramatic and historical status. Gangrenous ergotism (now known to be associated with consumption of contaminated cereals) was recorded as long ago as the ninth and tenth centuries. Many thousands of people died in post-war USSR from the mycotoxicosis—alimentary toxic aleukia—which was due to consumption of cereal moulded with *Fusaria* and highly contaminated with trichothecene toxins. It is noteworthy that, despite the high toxicity of many mycotoxins in laboratory animals, proof of human toxicity is extremely difficult to establish; this further emphasises the need for new approaches to toxicity assessment. The following examples illustrate this point.

Ochratoxin A, a potent animal nephrotoxin that can be found in cereals, has long been associated with endemic Balkan nephropathy, which is occasionally fatal. However, the association has been neither proved nor assigned to other agents. Likewise, the causal agents of human oesophageal cancer, with high incidence in Transkei and Iran, are unknown. Strong suspicions centre on the role of toxins from certain species of *Fusaria*. The role of aflatoxin B in human disease remains especially controversial. This mycotoxin is the most potent carcinogen tested in certain animals, yet only recently has the International Agency for Research on Cancer accepted that it is probably a human carcinogen,[14] a decision disputed by some workers.[15] The other toxic effects of aflatoxin B in man are equally unclear. Even though analytical methodology is adequate and sufficient materials are available for research, knowledge of potential human toxicity remains poor and is often dependent on epidemiological surveys and very little else.

Factors that affect dietary exposure

The following factors would contribute to substantial increases in dietary exposures to a wide range of biologically-active natural substances. (1) Nutritional advice and recommendations to increase consumption of green vegetables, vegetable protein, and "fibre". Though correct, such advice will lead to an increase in exposure of various natural chemicals, whose long-term health effects are unknown. (2) Dietary trends and fashions—eg, vegetarianism and consumption of exotic or ethnic foods, or of supposedly beneficial products, such as potato peel. (3) Increased public interest in health foods, herbal remedies, and herbal "medicines". Many of these products contain the highly toxic pyrrolizidine alkaloids (see R. J. Huxtable in ref 8). Since many herbal remedies were previously used for certain disorders, exposures were low and discontinuous. Many of these remedies are now being recommended as tonics and "pick-me-ups", so exposures are very much larger and more regular, if not continuous. Carlsson[16] has recently recorded the association between liver abnormalities and herbal preparations. Huxtable's advice to avoid consumption of anything that contains comfrey should certainly be followed. (4) Current agricultural practices. During the next 10 years, consumer concerns about environmental and food safety issues will doubtless lead to a substantial decrease in the overall use of chemicals in agriculture. To maintain crop viability in the face of such a trend, plant pathologists and genetic engineers are developing new varieties with enhanced natural resistance. Disease resistance is complex, and many of the compounds that we describe have a role in plant defence strategies.[17] That an increase in endogenous disease resistance can be associated with increased concentrations of natural toxicants is clearly shown by the development of improved varieties of celery and potato, which have high enough levels of psoralens and glycoalkaloids, respectively, to pose a substantial risk to public health.[18]

Risk reduction

How can risk associated with consumption of a certain foodstuff be reduced or removed? First, elimination of the food from the diet is the only sure solution. However, it cannot be easily enforced and depends on the availability of alternative produce. It is not unusual for individuals to exclude certain dietary items. For example, people with coeliac disease avoid foods that contain gluten; migraine sufferers avoid chocolate, bananas, red wine, and other foods that trigger the symptoms; and inviduals who are deficient in lactase avoid milk and dairy products. Secondly, diet can be used to prevent or regulate disease; although this has many supporters at present, very little is known about the long-term effects or side-effects of the dietary chemicals concerned. Thirdly, the importance of processing—ie, washing, cooking, and fermentation—has long been recognised to detoxify food. However, care must be taken to ensure that the introduction of new technology or even minor modifications to existing processing conditions do not adversely affect the efficiency of detoxification; changes in culinary practice in the home may also have unforeseen consequences. Pulses and grain legumes contain highly toxic lectins (haemagglutinins), but the effects of these in man are negligible because boiling (which normally precedes consumption) inactivates the lectin molecules. The tendency, noted within the past decade among certain sections of the population, to only partly cook pulses—especially red kidney beans—and to serve them in salads has led to numerous cases of gastrointestinal disturbance. Many packets on sale now carry advice about proper cooking. Finally, with the possible exception of potato glycoalkaloids, plant breeders have paid little attention to natural toxicants and food safety. In view of greater public and social awareness of plant-related hazards this situation will certainly change. However, since many of the toxicants are important in plant protection, their removal or substantial reduction will not necessarily be straightforward.

References

1. Wodicka VO. *Food Chem News* 1971; **12**: 11–17.
2. Hall RL. Information, confidence and sanity in the food sciences. *Flavour Industry* 1971: 455–59.
3. British Medical Association Guide. Living with risk. Chichester: John Wiley and Sons, 1987: 24–27.
4. Miller K, ed. Toxicological aspects of food. London: Elsevier Applied Science, 1987.
5. Concon JM. Food toxicology. New York: Marcel Dekker, 1988.

6. Friedman M, ed. Nutritional and toxicological aspects of food safety. New York: Plenum, 1984.
7. Rechcigl M, Jr. CRC handbook of naturally-occurring food toxicants. Boca Raton: CRC Press, 1983.
8. Cheeke R, ed. Toxicants of plant origin. Boca Raton: CRC Press, 1989.
9. Denning DW. Aflatoxin and human disease. *Adverse Drug Reactions and Acute Poisoning Rev* 1987; **4**: 175.
10. Infante ME, Perez AM, Simao MR, et al. Outbreak of acute toxic psychosis attributed to *Mucuna pruriens. Lancet* 1990; **336**: 1129.
11. McMillan M, Thompson JG. An outbreak of suspected solanine poisoning in schoolboys. *Q J Med* 1979; **48**: 227–43.
12. Gee JM, Price KR, Ridout CL, Johnson IT, Fenwick GR. Effects of some purified saponins on transmural potential difference in mammalian small intestine. *Toxicol in vitro* 1989; **3**: 85.
13. Perl TM, Bédfard L, Kosatsky T, Hockin JC, Todd E, Remis RS. An outbreak of toxic encephalopathy caused by eating mussels contaminated with domoic acid. *N Engl J Med* 1990; **322**: 1775–80.
14. IARC Monographs on the evaluation of the carcinogenic risk of chemicals to humans (suppl 7). Lyon: International Agency for Research on Cancer, 1987: 83.
15. Stoloff L. Carcinogenicity of aflatoxins. *Science* 1987; **237**: 1283–84.
16. Carlsson C. Herbs and hepatitis. *Lancet* 1990; **336**: 1068.
17. Fenwick GR, Johnson IT, Hedley CL. Toxicity of disease-resistant plant strains. *Trends Food Sci Technol* 1990; **1**: 23–25.
18. Ames BN, Profet M, Gold LS. Pesticides (99·99% all natural), cell proliferation, and animal cancer tests. *Science* (in press).

17
Bovine spongiform encephalopathy

J. G. Collee

Bovine spongiform encephalopathy (BSE), a recently recognised disease of cattle, is caused by a transmissible agent that closely resembles, or is identical with, the transmissible agent of scrapie in sheep.[1] In this review I shall examine current anxieties about the possible presence of the BSE agent in our food.[2,3] Since our understanding of the disease and its transmissibility in cattle must await observations that will be made over some years to come, it is important to keep a reasonable perspective and to ensure that any speculative comment is consistent with what we already know. In much of our risk assessment in such circumstances, we are driven to look for persuasive parallels when directly relevant information is not available. Thus, it is necessary to review the following facts in any consideration of the possible oral transmission of BSE to man.

Spongiform encephalopathies of animals

SCRAPIE

Scrapie has been recognised as a disease of sheep for more than two centuries.[4] The disease also affects goats. Since the agent of scrapie (or of BSE) has not been isolated, present descriptions are based on observations from transmission experiments with infected tissue preparations that have had various specified treatments. Scrapie-like agents are not recognisable as bacteria, and they differ from viruses—ie, they are highly resistant to heat and irradiation, and to other agents, such as formaldehyde and glutaraldehyde. Boiling or standard autoclaving (121°C for 15 min) does not inactivate these agents with certainty,[5] but exposure to 2% sodium hypochlorite or autoclaving in a porous load system at 134–138°C for 18 min achieves inactivation. Brain homogenate protects some viruses against inactivation by heat or chemicals.[5] Since the scrapie agent cannot be cultured, brain homogenates have had to be used for resistance tests; the resultant data are not likely to understate the agent's resistance but they may overstate it for circumstances in which the agent is not so protected.

Scrapie is naturally transmitted from ewe to lamb and horizontally between sheep, but opinions differ about modes of transmission. Pattison and colleagues[6,7] postulated that the eating of infected placentas may be the mechanism of apparent vertical transmission. The conflicting considerations that make it difficult to be sure about the oral route of infection as a natural pathway for the spongiform encephalopathies have been discussed by Gibbs et al;[8] scrapie has been experimentally transmitted to sheep and goats by the oral route, but in some cases the challenge doses were very large. The scrapie agent can be transmitted from affected sheep to mice or hamsters by intracerebral, intraperitoneal, or other injection of infected tissue preparations. Since there is experimental evidence that there are different scrapie strains and that mutation can occur, the presence of a genome is assumed.[9] Perhaps a lipid component gives some protection; the occurrence of a short single-stranded nucleic acid genome might be more compatible with such a daunting survival record, but this is speculative. The length of the incubation period in mice varies not only with the strain of scrapie agent, but also with the dose and route of administration; there is also host control exerted by a gene that affects the rate of replication of the agent in the central nervous system.[10]

That the agent has not been seen under high-resolution electronmicroscopy suggests that it is very small. The agent cannot be cultured in media or in cell cultures; its presence cannot be detected by tests for a known antigen; and it does not provoke the development of a specific antibody in the infected animal. The diagnosis presently relies on the clinical appearance of the affected animal and on the histopathology of the affected brain, which shows microscopic vacuoles in the grey matter (like a cross-section of a sponge) with loss of neurons. Electronmicroscopy of processed fractions of scrapie-infected brain reveals typical fibrils which have also been demonstrated in brain fractions from Creutzfeldt-Jakob disease in man.[11] These scrapie-associated fibrils (SAF) seem to be associated with infectivity and contain a specific host-coded glycoprotein, PrP_{27-30}, which is resistant to digestion by proteinase.[12] SAF may be a form of the infectious agents of these diseases, or they may be a unique pathological host response seen in these diseases.[11,12]

OTHER SPONGIFORM ENCEPHALOPATHIES

Apparently related wasting or encephalopathic syndromes have been noted in members of the deer family. A spongiform encephalopathy affecting mink on mink farms in the USA has been attributed to the feeding of infected sheep offal to mink, but the mode of transmission may be by inoculation from biting or scratching when feeding. Up to Feb 1, 1991, disease resembling BSE was also described in six captive exotic bovids (exotic ruminants) in England under circumstances that point to the same source as that for cattle (see below).[13] Spongiform encephalopathy has also been confirmed in twelve cats in Britain.

Spongiform encephalopathies of man

CREUTZFELDT-JAKOB DISEASE (CJD)

CJD is a spongiform encephalopathy of man,[14] but there is no evidence that the disease is an expression of natural transmission of scrapie to man; there is fairly strong circumstantial evidence that there is no direct causal association between scrapie in sheep and CJD in human beings.[15] A variant of CJD in man is called the Gerstmann-Sträussler syndrome, which is usually familial. CJD has been accidentally transmitted to patients by the injection of pituitary extracts or the transplantation of tissues from infected donors, or by the implantation of contaminated electrodes, but oral transmission of CJD has not been recognised in man. However, the oral transmission of kuru (see below), CJD, and scrapie to non-human primates has been achieved experimentally by workers who then believed that there could have been a link between scrapie and CJD in man.[8] It is difficult to resolve this view with the large amount of available epidemiological evidence that challenges it.

KURU

Kuru is an encephalopathy that affected a tribe in the highlands of New Guinea where it is said that cannibalism was practised until 1957. That women were predominantly affected was attributed to the reported custom that they ate the brains and other offal while the men ate the muscle. Taylor[16] has interpreted these reports more cautiously and calls attention to the possibility that kuru may not have been transmitted by the oral route. There is also a possibility that kuru and CJD are the same disease and that kuru was introduced into the tribe when a CJD-infected person was the subject of the mourning ritual.

Bovine spongiform encephalopathy

The cerebral pathology of the bovine disease was first described in 1987 in the UK by Wells and colleagues;[1] they found bilaterally symmetrical degenerative changes in the brainstem with discrete ovoid and spherical vacuoles (10–20 μm) or microcavities in the neuropil. Larger intracytoplasmic vacuoles (20–40 μm) were seen in some brainstem nuclei distending neurons to balloon forms with narrow rims of cytoplasm. Mild gliosis was sometimes noted, and sparse perivascular mononuclear cell infiltration in the brainstem parenchyma was seen in some cases; the virtual absence of an inflammatory reaction is a feature of the disease. Fibrils which resemble SAF can be demonstrated by electronmicroscopy of processed fractions of BSE-infected brain, and the specific glycoprotein PrP_{27-30} can be detected immunologically.

It is currently believed that BSE, first recognised in cattle in Britain in 1985–86, is likely to represent an unfortunate transmission of the scrapie agent from processed sheep protein or protein from subclinically affected cattle by the oral route. The incubation period in cattle seems to range from 2 to 8 years or more. Until the recognition of BSE in cattle and the postulated link with scrapie-infected sheep protein, most microbiologists would have discounted the likelihood of the oral transmission of scrapie across an animal species barrier that has apparently been so exclusive for so long. There is some evidence of a genetic predisposition in cattle that go down with BSE.[17]

It is of course reasonable to postulate that BSE might have developed a priori in the cow. Since cows and sheep share common pasturage and may often be in close contact, it is possible to develop several theories that could support or challenge such a view. The time-course of the events in Britain, with the introduction of protein supplements and the changes in rendering practices (see below), seems to support the generally held view that, whether or not BSE is primarily a disease of the cow, the feeding of commercial protein concentrates derived from ruminants was the important factor that precipitated the outbreak on a national scale.

It is possible that BSE may be occurring elsewhere in the world and that it may have occurred in Britain for some time before the first British cases were identified. There have been cases of the so-called "British" disease in the Channel Islands, Eire, and Oman where there have been clear links with British livestock producers and good veterinary services to make the diagnosis. It has been suggested that sporadic cases of BSE may have occurred in the USA where cows with paralytic disease are known as "downer cows";[18] the downer syndrome encompasses a range of neurological diseases. The first case of BSE in an animal born outside the British Isles was recorded in November, 1990. The disease was diagnosed in a 6-year-old Holstein cross cow in Switzerland, but the case seems to have no connection with the British outbreak.[19]

The rendering processes, in which discarded animal products are cooked and fats and proteins are recovered, were reviewed by the Working Party on Bovine Spongiform Encephalopathy (Southwood report),[20] which was commissioned by the Ministry of Agriculture, Fisheries and Food (MAFF), UK, to advise about what steps should be taken to halt the epidemic of BSE and to safeguard human health. The materials processed at rendering plants were from cattle, sheep, and pigs, and were also of "mixed origin including poultry". Various systems were described in the report, with a wide range of operating temperatures and times of exposure.

Some of the substantial changes in commercial rendering processes for ruminant offal in the late 1970s and early 1980s in England and Wales are believed to be associated with the use of a range of processing temperatures and the omission of various solvent steps as the markets for tallow declined. Lower sterilisation temperatures for the preparation of cattle feed were introduced in the 1980s. During that period, the numbers of sheep in Britain increased substantially and it is likely that many animals that were rejected for human consumption were rendered and processed to make protein for animal food supplements. The circumstances were very well reviewed by Wilesmith et al[21] in 1988. As a result of the Southwood report, changes in rendering processes now include a preference for continuous rather than batch processing.

The occurrence of BSE and the persuasive evidence that the disease is attributable to the practice of feeding ruminant-derived materials to herbivores obliged the British Government to take very prompt steps to reduce to a minimum the risk of a similar transmission from bovine material to man. The possibility of such a transmission must be considered against the lack of evidence of any such transmission of scrapie to man despite the consumption of sheep products in many forms for more than 250 years. Nonetheless, positive precautions were taken. BSE became a notifiable disease in the UK; animals suspected of having BSE and milk from BSE-infected cows are destroyed. A compensation assurance given to farmers in Great Britain in August, 1988, was considered to be inadequate and it has since been improved. The feeding of ruminant-derived protein to ruminants, including cattle, was banned in the UK in July, 1988, and the sale for human consumption of bovine offal products that might be regarded as likely replication sites for the BSE agent (brain, spleen, spinal cord, thymus, intestines, and tonsil, from animals older than 6 months) has been prohibited since November, 1989, in England and Wales,[22] and since Jan 31, 1990, in Scotland and Northern Ireland. The reasons for the exclusion of young calves from the regulations are that there is so far no evidence of vertical transmission of BSE and the minimum incubation period of the disease in cows seems to be about 2 years, with the peak incidence at 4 years. With the caution that must apply to observations about a disease with a long incubation period, the weight of evidence about scrapie indicates that it is unusual for infectivity to be associated with any tissue of scrapie-exposed lambs up to 10 months of age. However, our inability to demonstrate the presence of an elusive infective agent in a tissue cannot be interpreted as an absolute assurance of its absence from that tissue.

Risk assessment

BSE seems to be a new disease, and there is an obligation to assess its possible dangers to man. In practice, it is necessary to temper an extremely cautious approach based on a scientific awareness of the probability, however small, that anything can happen, with practical action based on probably reasonable odds that what we do is safe.

Does BSE pose a food hazard for man? How can we assess the risk? The safety of beef for man has not been tested and may not be testable.[3] However, the Southwood report[20] concluded that the evidence available in the 1980s indicated that cattle will prove to be a "dead-end host" for the disease agent and that it is most unlikely that BSE will have any implications for human health. The working party predicted that, with certain reservations, there might be a fall in the number of new cases of BSE in cows by about 1993, with disappearance of the disease from Great Britain by the end of the decade if the only mode of infection was by contaminated feed. By Feb 1, 1991, the cumulative total of confirmed cases of BSE in Britain was 23 346 on 10 437 farms. The features had seemed to be those of an extended common source outbreak in which all of the affected animals were index cases. However, it seems possible that secondary cases are occurring: the infection could have been recycled by the incorporation of infected bovine and ovine material from rendering factories into protein supplements before the ban came into operation. Thus, a bovine-adapted strain of the transmissible agent might have been given the opportunity to spread more widely. The numbers of cattle infected are likely to exceed the earlier predictions and the fall in the number of new cases may be delayed for some years until the ban is effective; this assumes that vertical or lateral transmission does not naturally occur.

The breakthrough of scrapie from sheep to cattle, if this is indeed what BSE represents, was probably assisted by several factors. It is likely that the feeding of sheep protein derived from infected sheep offal carried high challenge doses of the scrapie agent to the cow. It is not known whether the agent is transmitted to the cow by the oral route through an entirely undamaged gastrointestinal tract, whether abrasions in the tract (or elsewhere on the animal) may be important points of entry, or whether the conjunctival route or other routes should be considered. There is no evidence yet of either cow-to-cow (lateral) transmission or cow-to-calf (vertical) transmission of BSE.

The BSE agent could possibly be a variant of the scrapie agent with properties that have made it pathogenic for the cow (bovine-adapted). When this has happened in other models of infectious disease, the successful variant may be highly virulent for the new host but is not invariably adapted to cross other species barriers until further variation occurs. For example, strenuous efforts to transmit transmissible mink encephalopathy to mice were unsuccessful,[23] although transmission to hamsters was achieved.[24] However, the experimental dietary transmission of BSE to mice was recorded in 1990.[25] The mice were fed infected bovine brain and cerebrospinal fluid during 8 nights and days. This model may allow assessments of the relative infectivity of various bovine tissues (for mice). The experimental transmission of BSE to one of eight severely challenged pigs has been reported,[26] but this experiment bears no relation to the hazard for the pig of natural dietary exposure to bovine products. Nevertheless, the prohibition on the use of specified bovine offal or derived animal protein in animal feeds has now been widely extended to animals and poultry. It is noteworthy that the ban embraces pet food.

Taylor's assessment[16] of the risk of BSE to man is highly authoritative and carefully argued. He calls attention to the relative inefficiency of the oral route of challenge and to the very high doses used in experimental oral challenge tests with the transmissible agents of kuru and scrapie. Moreover, he points out that no association has yet been found to link an occupational exposure to potentially scrapie-infected tissues with any transmissible human degenerative encephalopahy in shepherds, abattoir workers, or butchers. By contrast, the iatrogenic transmission of the agent of Creutzfeldt-Jakob disease is well recognised[27] and strict precautions are now taken to protect patients and associated clinical workers from the hazard of inoculation or implantation of CJD-infected material. These two lines of thought provide added weight to the view that CJD is not a form of scrapie in man and that the emotive issues of kuru and CJD should not unduly influence our approach to BSE.

Exclusion systems to deal with the threat of BSE to man have been set up by the British Government and MAFF in response to the Southwood report and to the recommendations of other experts; the interim report of the Committee on Research into Spongiform Encephalopathies (Tyrrell report)[28] has indicated how we should proceed to get more relevant information to supplement the surveillance systems that are already in place. Some workers have argued that the British Government should have taken more draconian steps more quickly. The suggestion that calves of infected cows should be excluded from the chain until the risk of vertical transmission has been excluded by better scientific evidence presupposes that the parentage of all calves is reliably recorded and that all offspring of an affected cow can then be traced in our present system. A more feasible extra precaution would be to extend the ban on the sale of bovine offal products to include materials from young calves.[29] It has also been suggested in the lay press that any herd in which a case of BSE has occurred should be destroyed. Strongly emotive arguments are being advanced, but evidence to justify such extreme action is currently unavailable; the BSE scenario is very different from that of foot-and-mouth disease.[29]

Research initiatives

The Tyrrell report has identified key questions and has set priorities for research that should be done. A substantial amount of research is already in hand in the UK, Europe, Asia, and the USA to resolve many outstanding questions. The work includes detailed epidemiological studies, transmission experiments, genetic studies, and various investigations of the molecular biology of BSE. The transmission studies include work with mice, hamsters, sheep, goats, marmosets, pigs, chickens, mink, and cattle challenged by various routes. Special research projects have been set up to determine whether BSE is transmitted to calves in utero or in the immediate post-partum period, and whether an embryo from a BSE-affected donor can infect the recipient cow.

Prospect

Kimberlin[30] sets out the evidence against the view that CJD represents scrapie in man. If there is any direct link between the causes of scrapie and CJD, we must await the evidence that is presently lacking and we would have to set aside or explain the counter-evidence which is very substantial (see Taylor[16]). At present, it seems that an assuredly common factor in the transmissible dementias and spongiform encephalopathies is an aberrant form of prion protein.[31] We should not be alarmed if evolving immunocytochemical and molecular procedures demonstrate that we currently underestimate the incidence of prion disease in the population and that a substantial part of the formula has a genetic basis.[32] If the remarkable concept of prion disease gains ground quickly and is applied in practical diagnostic terms, Roberts and Gollinge[32] observed that any attempt to assess the impact of foodborne BSE on the incidence of transmissible dementias in man must take account of the improved rate of detection that would result from the new technology. At that stage, we can anticipate further pressures to control foodstuffs of bovine origin (whilst the aetiology may or may not have anything to do with the cow) and this is likely to continue until we have a specific marker for BSE.

The expert help of many colleagues is acknowledged. I especially thank Dr David M. Taylor, Dr Tony W. A. Little, Dr Ray Bradley, and Dr John Wilesmith for their guidance, but the views expressed are mine.

References

1. Wells GAH, Scott AC, Johnson CT, et al. A novel progressive spongiform encephalopathy in cattle. *Vet Rec* 1987; **121:** 419–20.
2. Holt TA, Phillips J. Bovine spongiform encephalopathy. *Br Med J* 1988; **296:** 1581–82.
3. Matthews WB. Bovine spongiform encephalopathy. *Br Med J* 1990; **300:** 412–13.
4. Kimberlin RH. Scrapie: the disease and the infectious agent. *Trends Neurosci* 1984; **7:** 312–16.
5. Rohwer RG. Virus-like sensitivity of the scrapie agent to heat inactivation. *Science* 1984; **223:** 600–02.
6. Pattison IH, Millson GC. Experimental transmission of scrapie to goats and sheep by the oral route. *J Comp Pathol* 1961; **71:** 171–76.
7. Pattison IH, Hoare MN, Jebbett JN, Watson WA. Spread of scrapie to sheep and goats by oral dosing with foetal membranes from scrapie-infected sheep. *Vet Rec* 1972; **9:** 465–68.
8. Gibbs CJ, Amyx HL, Bacote A, Masters CL, Gajdusek DC. Oral transmission of Kuru, Creutzfeldt-Jakob disease, and scrapie to nonhuman primates. *J Infect Dis* 1980; **142:** 205–08.
9. Bruce ME, Dickinson AG. Biological evidence that scrapie agent has an independent genome. *J Gen Virol* 1987; **68:** 79–89.
10. Kimberlin RH. Scrapie: how much do we really understand? *Neuropath Appl Neurobiol* 1986; **12:** 131–47.
11. Merz PA, Somerville RA, Wisniewski HM, Manuelidis L, Manuelidis EE. Scrapie-associated fibrils in Creutzfeldt-Jakob disease. *Nature* 1983; **306:** 474–76.
12. Diringer H, Gelderblom H, Hilmert H, Ozel M, Edelbluth C, Kimberlin RH. Scrapie infectivity, fibrils and low molecular weight protein. *Nature* 1983; **306:** 476–78.
13. Kirkwood JK, Wells GAH, Wilesmith JW, Cunningham AA, Jackson SI. Spongiform encephalopathy in an arabian oryx (*Oryx leucoryx*) and a greater kudu (*Tragelaphus strepsiceros*). *Vet Rec* 1990; **127:** 418–20.
14. Behan PO. Creutzfeldt-Jakob disease. *Br Med J* 1982; **284:** 1658–59.
15. Anon. BSE and scrapie: agents for change. *Lancet* 1988; ii: 607–08.
16. Taylor DM. Bovine spongiform encephalopathy and human health. *Vet Rec* 1989; **125:** 413–15.
17. Wijeratne WVS, Curnow RN. A study of the inheritance of susceptibility to bovine spongiform encephalopathy. *Vet Rec* 1990; **126:** 5–8.
18. Gibbs CJ, Safar J, Ceroni M, Di Martins A, Clark WW, Hourrigan JL. Experimental transmission of scrapie to cattle. *Lancet* 1990; **335:** 1275.
19. Anon. BSE case found on the Continent. *Vet Rec* 1990; **127:** 462.
20. Report of the Working Party on Bovine Spongiform Encephalopathy. London: Department of Health, 1989.
21. Wilesmith JW, Wells GAH, Cranwell MP, Ryan JBM. Bovine Spongiform Encephalopathy: epidemiological studies. *Vet Rec* 1988; **123:** 638–44.
22. Statutory instruments 1989 No. 2061. Food. The Bovine Offal (Prohibition) Regulations. London: HM Stationery Office. 1989: 11–12.
23. Taylor DM, Dickinson AG, Fraser H, Marsh RF. Evidence that transmissible mink

encephalopathy agent is biologically inactive in mice. *Neuropathol Appl Neurobiol* 1986; **12**: 207–15.

24. Kimberlin RH, Cole S, Walker CA. Transmissible mink encephalopathy (TME) in Chinese hamsters: identification of two strains of TME and comparisons with scrapie. *Neuropathol Appl Neurobiol* 1986; **12**: 197–206.

25. Barlow RM, Middlleton DJ. Dietary transmission of bovine spongiform encephalopathy to mice. *Vet Rec* 1990; **126**: 111–12.

26. Dawson M, Wells GAH, Parker BNJ, Scott AC. Primary parenteral transmission of bovine spongiform encephalopathy to the pig. *Vet Rec* 1990; **127**: 338.

27. Masters CL, Harris JO, Gajdusek C, et al. Creutzfeldt-Jakob disease: patterns of world-wide occurrence and the significance of familial and sporadic clustering. *Ann Neurol* 1979; **5**: 177–88.

28. Interim Report. Consultative Committee on Research into Spongiform Encephalopathies. Ministry of Agriculture, Fisheries and Food; and the Department of Health. London: MAFF publications, 1989.

29. Anon. BSE in perspective. *Lancet* 1990; **335**: 1252–53.

30. Kimberlin, RH. In: Parker MT, Collier LH, eds. Topley & Wilson's Principles of Bacteriology, Virology and Immunity, 8th ed, Vol 4 (Collier LH, Timbury MC, eds); London: Edward Arnold, 1990, 678–80.

31. Anon. Prion disease—spongiform encephalopathies unveiled. *Lancet* 1990, **336**: 21–22.

32. Roberts GW, Collinge J. Bovine spongiform encephalopathy. *Br Med J* 1990; **300**: 934–44.

Index